VOL. 4

Uncovering
Student Ideas
in Science

25 NEW Formative Assessment Probes

W9-AQS-459

VOL. 4

Uncovering
Student Ideas
in Science

25 NEW Formative Assessment Probes

By Page Keeley
and Joyce Tugel

NATIONAL SCIENCE TEACHERS ASSOCIATION

Arlington, Virginia

University Library of Columbus
4555 Central Avenue, LC 1600
Columbus, IN 47203-1892

National Science Teachers Association

Claire Reinburg, Director
Jennifer Horak, Managing Editor
Judy Cusick, Senior Editor
Andrew Cocke, Associate Editor
Betty Smith, Associate Editor

ART AND DESIGN
Will Thomas, Jr., Director
Cover, Inside Design, and Illustrations by Linda Olliver

PRINTING AND PRODUCTION
Catherine Lorrain, Director

NATIONAL SCIENCE TEACHERS ASSOCIATION
Francis Q. Eberle, PhD, Executive Director
David Beacom, Publisher

Copyright © 2009 by the National Science Teachers Association.
All rights reserved. Printed in the United States of America.
11 10 09   5 4 3 2

Library of Congress Cataloging-in-Publication Data

Keeley, Page.
Uncovering student ideas in science / by Page Keeley, Francis Eberle, and Lynn Farrin.
    v. cm.
Includes bibliographical references and index.
Contents: v. 1. 25 formative assessment probes
ISBN 0-87355-255-5
1. Science--Study and teaching. 2. Educational evaluation. I. Eberle, Francis. II. Farrin, Lynn.
III. Title.
Q181.K248 2005
507'.1--dc22
                        2005018770

NSTA is committed to publishing material that promotes the best in inquiry-based science education. However, conditions of actual use may vary, and the safety procedures and practices described in this book are intended to serve only as a guide. Additional precautionary measures may be required. NSTA and the authors do not warrant or represent that the procedures and practices in this book meet any safety code or standard of federal, state, or local regulations. NSTA and the authors disclaim any liability for personal injury or damage to property arising out of or relating to the use of this book, including any of the recommendations, instructions, or materials contained therein.

PERMISSIONS
You may photocopy, print, or e-mail up to five copies of an NSTA book chapter for personal use only; this does not include display or promotional use. Elementary, middle, and high school teachers *only* may reproduce a single NSTA book chapter for classroom or noncommercial, professional-development use only. For permission to photocopy or use material electronically from this NSTA Press book, please contact the Copyright Clearance Center (CCC) (*www.copyright.com; 978-750-8400*). Please access *www.nsta.org/permissions* for further information about NSTA's rights and permissions policies.

Contents

Physical Science and Unifying Themes Assessment Probes

Life, Earth, and Space Science Assessment Probes

. . . .

Dedication

This book is dedicated to Dr. Richard Konicek-Moran, professor emeritus at the University of Massachusetts-Amherst, for his 50 years of service to the science community in raising awareness of the importance of students' conceptions; to his wife, Kathleen, a botanist and botanical illustrator extraordinaire; and to Dr. Bob Barkman, our "Earth Is (Not) Flat" friend and colleague. Who would have predicted that a phone call made years ago to inquire about the "probes" would have led to such a rewarding partnership and cherished friendship among the authors and these three individuals who have inspired us in our work.

. . . .

Preface

This book is the fourth in the highly successful *Uncovering Student Ideas in Science* series. The addition of 25 more formative assessment probes has now expanded the collection to a total of 100 science elicitation questions that provide teachers with insights into student thinking seldom revealed through standard science assessment questions. In this book, a new addition to the Earth, space, physical, and life science probes is the inclusion of two probes that target important unifying themes in science models and systems. Collectively, these 100 probes focus on important fundamental ideas in science that cut across multiple grade spans.

Regardless of whether you teach elementary, middle, or high school science, students' preconceptions can be tenacious and often follow students from one grade span to the next. Taking the time to elicit and examine student thinking is one of the most effective ways to support instruction that leads to conceptual change and enduring understanding. It is also the starting point for differentiating instruction to meet the content needs of all students.

Since Volume 1 was released in October 2005, Volume 2 in 2007, and Volume 3 in 2008, thousands of K–12 teachers, university faculty, and professional developers have used these probes to bring to the surface the ideas students and teachers have that they might not even be aware of. The response to these probes

has been very encouraging. Teachers have frequently remarked to us that they now know much more about their students and student learning. They also report that the probes have significantly changed their instruction as well as their classroom environments. Teachers spend more time letting their students do the talking, listening carefully to their ideas, and constantly thinking about the next steps they need to take to move their students from where they are to where they need to go in order to develop conceptual understanding. Old habits, such as the need to grade every piece of student work or acknowledge only right answers, have changed to allow students the opportunity to express their thinking safely—that is, in classroom cultures that welcome and value new ideas.

Not only are teachers using the probes to elicit students' ideas and inform their instructional practices, but teachers are also using the probes to transform their own learning. In our work at the Maine Mathematics and Science Alliance, we provide professional development to many school districts, math-science partnership projects, and other teacher enhancement initiatives throughout the United States that use these probes in their teacher professional development programs. The insights we have gained from working with teachers show that the probes have challenged teachers' own thinking about ideas in science, brought to the

Preface

surface long-held misconceptions that many teachers were unaware they held, and revealed how some instructional activities and methods can lead to reinforcing common misconceptions without teachers realizing it. In addition, the Teacher Notes that follow each probe have increased teachers' ability to see the link among key ideas in the standards, developmentally appropriate instruction, students' commonly held ideas, and strategies for addressing students' ideas. All of this information gained from using the probes has led to profound changes in teachers' content knowledge, pedagogy, and beliefs about how students learn science.

In Volume 4, we decided to focus on ways to balance formative assessments with summative assessments (e.g., classroom-based, district, and state assessments) because of the widespread interest in this balancing challenge. We believe it is important to distinguish between these two types of assessment, recognize the link between them, and stay true to the purposes of each.

As the interest in formative assessment has skyrocketed and has become more prominent in local, state, and national efforts to improve science learning, the term *formative assessment* is being "hijacked" in the name of more practice for test taking. Publishers market sets of drill questions to prepare students for standardized tests and call them formative assessments. These questions are nothing more than a wolf in sheep's clothing. You can dress the wolf up in a sheepskin so it looks like a sheep, but underneath it still behaves like a wolf. Likewise, you can package test preparation questions as

"formative assessments," but underneath they are nothing more than questions limited in scope and depth that diminish quality instructional time and do little to promote learning and enduring understanding.

While you are probably most interested in using the 25 probes provided in this book, don't overlook the Introduction (pp 1–8) or the introductions in Volumes 1–3. Each introduction will expand your understanding of formative assessment and its inextricable link to instruction and learning. Volume 1 gives an overview of formative assessment. It also provides background on probes as specific types of formative assessments and how they are developed. Volume 2 describes the link between formative assessment and instruction and suggests ways to embed the probes into your teaching. Volume 3 describes how you can use the probes and student work to deepen your understanding of teaching and learning. This volume (Volume 4) describes the relationship between formative assessment and summative assessment. Collectively, the introductions in all four volumes will increase your assessment literacy and instructional repertoire. In addition, they will deepen your understanding of effective science teaching and learning.

The Teacher Notes that accompany each probe are made up of the following 10 elements.

Purpose

This section describes the general concept or topic targeted by the probe and the specific idea that is being elicited. It is important to be clear as to what the probe is going to reveal.

Being clear about the purpose of the probe will help you decide if the probe fits your intended learning target.

Related Concepts

Each probe is designed to target one or more related concepts that cut across grade spans. These concepts are described in the Teacher Notes and are also included on the matrix charts on pages 10 and 90. A single concept may be addressed by multiple probes. You may find it useful to use a cluster of probes to target a concept or specific ideas within a concept. For example, there are three probes in this volume that target the concept of natural selection.

Explanation

A brief scientific explanation, reviewed by scientists and content specialists, accompanies each probe and provides clarification of the scientific content that underlies the probe. The explanations are designed to help you identify acceptable or "best" answers (sometimes there is no "right" answer) and to clarify any misunderstandings you might have about the content. The explanations are not intended to provide detailed background knowledge on the concept, but they do provide enough explanation to connect the idea(s) in the probe with the science concept it is based on. If you need further explanation of the content, the Teacher Notes also list NSTA resources, such as the series *Stop Faking It! Finally Understanding Science So You Can Teach It* or Science Objects in the NSTA Learning Center, that will enhance and extend your understanding of the content.

Curricular and Instructional Considerations

The probes in this book do not target a single grade level as summative assessments do. Rather, they provide insights into the knowledge and thinking your own students may have regarding a topic as they developmentally progress or move from one grade span to the next. Some of the probes can be used in grades K–12 while others may cross over just a few grade levels. Teachers from two grade spans (e.g., elementary and middle school) might decide to use the same probe and come together and discuss their findings. To do this it is helpful to have insight into what students typically experience at a given grade span as it relates to the ideas elicited by the probe. Because the probes do not prescribe a specific grade level for use, you are encouraged to read the curricular and instructional considerations and decide if your students have had sufficient experience and the readiness to make the probe useful.

The Teacher Notes also describe how the information gleaned from the probe is useful at a given grade span. For example, it might be useful for planning instruction when an idea in the probe is a grade-level expectation or it might be useful at a later grade to find out whether students have sufficient prior knowledge to move on to the next level. Sometimes the student learning data gained through use of the probe indicate that you might have to back up several grade levels to teach ideas that are not really clear to students.

We deliberately chose not to suggest a grade level for each probe. If the probes were

Preface

intended to be used for summative purposes, a grade level, aligned with a standard, would be suggested. However, these probes have a different purpose. Do you want to know more about the ideas your students are expected to learn in your grade-level standards? Are you interested in how preconceived ideas develop and change across multiple grade levels in your school, sometimes even before they are formally taught? Are you interested in whether students have acquired a scientific understanding of previous grade-level ideas before you introduce higher-level concepts? The descriptions of grade-level considerations in this section can be coupled with the section that lists related ideas in the national standards in order to make the best judgment about grade-level use.

Administering the Probe

In this section, we suggest ways to administer the probe to students, including a variety of modifications that may make the probe more useful at certain grade spans. For example, we might recommend eliminating certain examples from a justified list for younger students who may not be familiar with particular words or examples or adding more sophisticated examples for older students. The notes also include suggestions for demonstrating the probe context with artifacts or ways to elicit the probe responses while students interact within a group. This section often refers to techniques described in *Science Formative Assessment: 75 Practical Strategies for Linking Assessment, Instruction, and Learning* (Keeley 2008) that

move the probes beyond paper-and-pencil tasks to interactive classroom strategies.

Related Ideas in the National Standards

This section lists the learning goals stated in the two national documents generally considered the "national standards": *Benchmarks for Science Literacy* (AAAS 1993) and *National Science Education Standards* (NRC 1996). Because the probes are not designed as summative assessments, the learning goals listed from these two documents are not intended to be considered as alignments but rather as related ideas connected to the probe. Some targeted ideas, such as a student's conception of the difference between weight and pressure, as seen in the probe "Standing on One Foot" on page 61, are not explicitly stated as learning goals in the standards but are clearly related to national standards concepts that address specific ideas about forces. When the ideas elicited by a probe appear to be a strong match with a national standard's learning goal, these matches are indicated by a star symbol (★).

Related Research

Each probe is informed by related research when it is readily available. Because the probes were not designed primarily for research purposes, an exhaustive literature search was not conducted as part of the development process. We drew primarily on three comprehensive research summaries commonly available to educators: Chapter 15 in *Benchmarks for Science Literacy* (AAAS 1993), *Making Sense of Secondary Science: Research Into Children's*

Ideas (Driver at al. 1994), and the research notes in the *Atlas of Science Literacy, Volume 2* (AAAS 2007). Although the first two resources describe studies that have been conducted in past decades and involved children not only in the United States but in other countries as well, many of the results of these studies are considered timeless and universal. Many of the ideas students held that were uncovered in the 1980s and 1990s research still apply today.

It is important to recognize that geography and cultural and societal contexts can influence students' thinking, but research also indicates that many of the ideas students have are pervasive regardless of geographic boundaries and societal and cultural influences. Hence the descriptions from the research can help you better understand the intent of the probe and the variety of responses your students are likely to have. As you use the probes, you are encouraged to seek new and additional research findings. One source of updated research can be found on the Curriculum Topic Study (CTS) website at *www.curriculumtopicstudy.org*. A searchable database on this site links each of the CTS topics to additional research articles and resources.

Suggestions for Instruction and Assessment

After analyzing your students' responses, it is up to you to decide on appropriate interventions and instructional strategies for your students. We have included suggestions gathered from the wisdom of teachers, the knowledge base on effective science teaching, and our own collective experience as former teachers and specialists involved in science education. These are not exhaustive or prescribed lists but rather suggestions that may help you modify your curriculum or instruction in order to help students learn ideas that they may be struggling with. It may be as simple as realizing that you need to be careful how you use a particular word in science. Learning is a very complex process and most likely no single suggestion will help all students learn the science ideas. But that is part of what formative assessment encourages—thinking carefully about the variety of instructional strategies and experiences needed to help students learn scientific ideas. As you become more familiar with the ideas your students have and the multifaceted factors that may have contributed to their misunderstandings, you will identify additional strategies that you can use to teach for conceptual change.

Related NSTA Science Store Publications, NSTA Journal Articles, NSTA SciGuides, NSTA SciPacks, and NSTA Science Objects

NSTA's journals, books, SciGuides, SciPacks, and Science Objects are increasingly targeting the ideas students bring to their learning. We have provided suggestions for additional readings that complement or extend the use of the individual probes and the background information that accompanies them. For example, Bill Robertson's *Stop Faking It!* series of books may be helpful in clarifying concepts teachers struggle with. A journal article from one

Preface

of NSTA's elementary, middle school, or high school journals may provide additional insight into students' misconceptions or provide an example of an effective instructional strategy or activity that can be used to develop understanding of the ideas targeted by a probe. Other resources listed in this section provide a more comprehensive overview of the topic addressed by the probe.

Related Curriculum Topic Study Guides and References

NSTA is copublisher of the book *Science Curriculum Topic Study: Bridging the Gap Between Standards and Practice* (Keeley 2005). This book was developed as a professional development resource for teachers with funding from the National Science Foundation and is available through NSTA Press. It provides a set of 147 curriculum topic study (CTS) guides that can be used to learn more about a science topic's content, examine instructional implications, identify specific learning goals and scientific ideas, examine the research on student learning, consider connections to other topics, examine the coherency of ideas that build over time, and link understandings to state and district standards. The CTS guides use national standards and research in a systematic process that deepens teachers' understanding of the topics they teach.

The CTS guides that were used in the development of the probes in this book are listed before each reference list. Teachers who wish to delve deeper into the standards and research-based findings that were used to develop the

probes may wish to use the CTS guides for further information.

In addition, Chapter 4 in the CTS book describes the process for developing an assessment probe that links standards and research on learning. Teacher educators, assessment developers, and others who want to engage groups in developing their own assessment probes will find professional development materials in *A Leader's Guide to Science Curriculum Topic Study: Designs, Tools, and Resources for Professional Learning* (Mundry, Keeley, and Landel 2009).

References are provided for the standards and research findings cited in the Teacher Notes.

. . . .

We hope this fourth volume of probes will be as useful to you as the other three volumes. If the interest continues in the *Uncovering Student Ideas in Science* series, we will continue to produce new books and assessment tools. If there are particular ideas you would like to see targeted in future volumes of *Uncovering Student Ideas in Science*, please contact the primary author of the series, Page Keeley, at *pagekeeley@gmail.com* or *pkeeley@mmsa.org*. Beginning in the spring of 2009, visit the Uncovering Student Ideas website—http://*uncoveringstudentideas.org*—where the author shares new information and updates related to assessment probes and maintains a blog on formative assessment in science.

References

American Association for the Advancement of Science (AAAS). 1993. *Benchmarks for science literacy.* New York: Oxford University Press.

American Association for the Advancement of Science (AAAS). 2007. *Atlas of science literacy*, Vol. 2. Washington, DC: AAAS.

Driver, R., A. Squires, P. Rushworth, and V. Wood-Robinson. 1994. *Making sense of secondary science: Research into children's ideas*. London: RoutledgeFalmer.

Keeley, P. 2005. *Science curriculum topic study: Bridging the gap between standards and practice*. Thousand Oaks, CA: Corwin Press.

Keeley, P. 2008. *Science formative assessment: 75 practical strategies for linking assessment, instruction, and learning*. Thousand Oaks, CA: Corwin Press.

Mundry, S., P. Keeley, and C. Landel. 2009. *A leader's guide to science curriculum topic study: Designs, tools, and resources for professional learning*. Thousand Oaks, CA: Corwin Press.

National Research Council (NRC). 1996. *National science education standards*. Washington, DC: National Academy Press.

Acknowledgments

The assessment probes in this book have been extensively field-tested by several teachers and hundreds of students. We would like to thank the teachers and science coordinators we have worked with for their willingness to field-test probes, share student data, and contribute ideas for additional assessment probe development.

In particular we would like to acknowledge the following people for their willingness to field-test probes in this volume: Kendra Bausch, Huntington Beach, CA; Robin Bebo-Long, Proctorsville, VT; Andrew Bosworth, Monmouth, ME; Linda Carpenter, Lovell, ME; Beth Chagrasulis, Naples, ME; Marilyn Curtis, Lisbon, ME; Candy Darling, Portsmouth, NH; Barbara Fortier, Biddeford, ME; Cassie Greenwood, El Paso, TX; Sue Kistenmacher, Wiscasset, ME; Susan Klemmer, Rockport, ME; Mark Koenig, Gardiner, ME; Kristine Lawrence, Scarborough, ME; Kris Moniz, Cape Elizabeth, ME; Jacey Morrill, Cumberland, ME; Margaret Morton, South Bristol, ME; Michael O'Brien, Kennebunk, ME; Suzi Ring, Brunswick, ME; Jerline Robie, St. Louis, MO; and Ken Vencile, Rockport, ME. We sincerely apologize if we overlooked anyone.

We would also like to thank our colleagues at the Maine Mathematics and Science Alliance (MMSA) (*www.mmsa.org*), science specialists we have worked with in school districts throughout the United States, professional development colleagues, Math-Science Partnership directors, and university partners nationwide. We are deeply appreciative of the efforts of the National Science Teachers Association in supporting formative assessment. We are greatly appreciative of all the support and assistance provided through the outstanding staff of NSTA Press including David Beacom, Claire Reinburg, and Judy Cusick. And last but not least, special thanks go to our graphic artist, Linda Olliver.

About the Authors

Page Keeley is the primary author and assessment probe developer of the *Uncovering Student Ideas in Science* series. She is the senior science program director at the Maine Mathematics and Science Alliance, where she has worked since 1996 developing and leading various leadership and professional development initiatives and projects in New England and nationally. Page consults with school districts, math-science partnership programs, and various initiatives throughout the United States in the areas of formative assessment, leadership, curriculum topic study (CTS), coaching and mentoring, and conceptual change instruction and is a frequent speaker at national conferences. She has authored numerous books on formative assessment and curriculum leadership and served as principal investigator and project director of three National Science Foundation grants, *Science Curriculum Topic Study: A Systematic Approach to Utilizing National Standards and Cognitive Research; The Northern New England Co-Mentoring Network;* and *PRISMS: Phenomena and Representations for Instruction of Science in Middle Schools*. Page is a former high school and middle school teacher and received the Presidential Award for Excellence in Secondary Science Teaching in 1992 and a Milken National Educator Award in 1993. In addition, Page Keeley is the 2008–2009 president of the National Science Teachers Association.

Joyce Tugel is a coauthor of two books in the *Uncovering Student Ideas in Science* series. She has been a science specialist at the Maine Mathematics and Science Alliance since 2005, where she directs projects in professional development, conceptual change teaching, environmental literacy, and engineering design. She is the project director of four Maine State Math-Science Partnership grants and a National Oceanic and Atmospheric Administration–funded tri-state environmental literacy project called Earth as a System is Essential: Seasons and the Seas. Joyce consults with school districts and organizations throughout the United States in professional development related to science curriculum, instruction, and assessment. Joyce has worked as a high school chemistry teacher and as a researcher in microbial biogeochemistry. She received the Presidential Award for Excellence in Secondary Science Teaching in 1998, a Milken National Educator Award in 1999, and the New England Institute of Chemists Secondary Teaching Award in 1999. Joyce is a former NSTA Division Director for Professional Development.

Introduction

> When teachers are asked how they assess their students, they typically talk about tests, examinations, quizzes, and other formal methods. When they are asked how they know whether their students have learned what they have taught, the answers are very different.
>
> —Dylan Wiliam, *Assessing Science Learning* (2008, p. 6)

Our districtwide K–12 science team came back from a National Science Teachers Association conference last spring all fired up. We had gone to a session on formative assessment and learned about the *Uncovering Student Ideas in Science* series of assessment probes and a variety of science formative assessment classroom techniques (FACTs) we could use with the probes. The presenter started the session by asking us to call out the first word that came to mind when we heard the word *assessment*. In unison, most of the people in the room called out "testing!" As the presenter then pointed out, *assessment,* and particularly formative assessment, is not necessarily about testing; it is about what you can do to improve learning and, ultimately, get better test results.

During the session, we, as learners, used the probes and the FACTs and came to realize how powerful they are. We had heard the phrase *assessment **for** learning* before, and for

Introduction

the first time, we felt we really knew what it meant. The speaker talked about the importance of using assessments like the probes and FACTs to create a balanced system of assessment that provides useful information at the beginning of instruction to promote thinking and inform instruction. This "front-end" assessment leads to better results at the "back end," when students are tested on what they have learned.

This approach to assessment started to make sense to us. All the concepts and ideas from the practice tests our students were taking to prepare for the state assessment were quickly forgotten by the students, even when it came time to take the test. We knew drilling with sample test questions wasn't the best solution for raising test scores, but we had always hoped it would help a little. We now realized that maybe formative assessment was what we needed to do more of at the beginning and throughout a lesson so that students—rather than memorize a lot of discrete facts—would have an opportunity to confront and work through their ideas before taking a test. After all, science is different from other subjects—it's about ideas that explain the natural world and processes that help us make sense of that world.

Many of the people in the audience had been using the probes and shared their stories. We connected with several folks from districts like ours and got all kinds of good ideas about how to use these tools to improve student learning and teacher practice. Before we left the conference we went to the NSTA bookstore and bought copies of the *Uncovering Student Ideas in Science* series. We started reading them on the plane ride home. We couldn't stop talking about them and our ideas for using them! We felt we had finally found a solution to the struggles we face in our district in balancing accountability and reporting with students' opportunity to learn.

Teaching and assessing for understanding aren't about more teaching, more materials, and more testing; they are about more opportunities to learn by promoting thinking and bringing students' ideas to the surface. Here was a set of assessments already developed for us to use with links to the key ideas in the standards, descriptions of the research that the probe was based on, and suggestions for things to do in the classroom to help students learn. These assessments would save us months of work that might have been spent developing our own formative assessments.

When we got back to our district we shared what we had learned in our grade-level teams with teachers and administrators. We argued over what *formative assessment* really meant and whether our current practices were consistent with what the research describes as good assessment practice. We all decided we wanted to move beyond practice for test taking and deadly drill sessions. Our superintendent surprised us with her enthusiasm and her offer to provide funding to support stipends and copies of the series for after-school professional learning communities. These groups would come together and study this new assessment technique, try it out in the classroom, examine student results, and report back.

After several months of exploring the theory and practice of formative assessment in collaborative groups, the broad consensus of our learning communities was that formative assessment worked! Students were more engaged in learning science, they began to write more extensively and to converse scientifically about their ideas, and they took more ownership in the learning process. Their explanations became much richer and we knew better how to then tailor learning to address the students where they were, rather than where our textbook and pacing guide said they should be. Our classroom questioning had changed from an ongoing monologue back and forth between the teacher and students to one in which there was rich dialogue among students working together with guidance from us in resolving their ideas. All students were involved and felt safe to share their thinking.

As we tried out formative assessment, we also looked at our summative assessments. We quickly found that released items from prior state assessments, our district-developed standardized tests, and even our own classroom tests did not give us the kind of data we needed to know exactly what the students' learning problems were. We knew our students weren't doing well in some areas such as matter and energy interactions or Earth systems, but we didn't know exactly what the learning problems were until we used the formative assessment probes as well as other new teaching strategies that probe students' thinking.

We decided to match up the probes with the assessment reporting categories aligned to our state standards, administer the probes across grade levels, analyze the results, and match the results to our district test data. Lo and behold, we found that the problem areas for tenth graders weren't much different from the problems of our middle school and elementary students.

Good data about student learning are at the core of improvement in student achievement. We began to use the Collaborative Inquiry into Examining Student Thinking (CIEST) protocol to look at student work from the probes (Mundry, Keeley, and Landel 2009). The probes revealed not only information we could use to inform our instruction but also a lot about the gaps in our K–12 curriculum that were affecting learning from one grade level to the next. No wonder certain problems continued after the fourth-grade test and the eighth-grade test. As a district, we had never collected the rich kind of formative data that could be used to pinpoint what the learning problems were, how they originated,

Introduction

and why they persisted from one grade to the next. Instead of looking at test scores—a single snapshot in time—and saying we needed to reteach the same material, we could now see just what we needed to focus on better. And just as important, our students began to experience the conceptual change that happens when they realize that their preconceptions no longer make sense to them and they start to construct new explanations.

Now we use the probes to try to improve the quality of our district assessments, and we look for evidence that students understand key ideas in science that may have been riddled with misconceptions in the past. We match our summative data to our formative data and identify patterns and discrepancies. *Accountability* isn't such a scary word anymore. We now have the right tools and processes to take the guesswork out of assessment and ensure that our students are ready to "show what they know."

As for the future, although we realize that not all of our students are going to leave our district with plans to be scientists, the chances are now much better that they will leave being science literate, ready to use their knowledge and skills to understand real-world issues and problems that require an understanding of the basic principles of science.

The above vignette is a composite account, drawn from the many stories we have heard from science educators who are using the *Uncovering Student Ideas in Science* series. It shows how formative assessment can provide valuable information to teachers and students to promote learning and inform instruction, creating a more balanced system of assessment that does not overly rely on summative assessment. Assessment isn't only about testing, and this book is not about assessments that are graded and then used to pass judgment on students about the extent to which they have achieved a learning goal. This book is about using a type of assessment, called a *probe*, for diagnostic and formative purposes.

The probes are selected-response items specifically designed to reveal student misconceptions that the research literature on student learning has documented. Throughout this book and the other three books in the series, we use the word *misconception* in a general way to refer to the ideas students bring to their learning that are not yet fully formed and not scientifically correct at the level we would expect. Other words to describe students' ideas include *preconceptions, naive ideas, partially formed ideas, facets of understanding,* and *alternative conceptions.* Although teachers tend to use the word *misconception* in a pejorative way to describe students' ideas that are not the same as the scientific ideas we want them to understand, misconceptions can be useful if we use them to build a bridge between where students are in their thinking and where we eventually want them to be.

Introduction

The assessment probes in this book are designed to help teachers build that bridge. The bridge begins with finding out what students think about important ideas in science that they will use throughout their learning. Although the probes were written primarily to target a K–12 student audience, they can be used with adult learners as well, including university students and science teachers participating in professional development. The previous three volumes provided background information for teachers on formative assessment. We encourage you to collect all the books in the series and read the introductory material to expand your understanding of assessment and its role in teaching, learning, and professional development.

In addition to the 25 probes in each volume, Volume 1 contains an overview of formative assessment—what it is and how it is used. Volume 2 introduces ways to integrate assessment with instruction, and Volume 3 provides an introduction to using the probes, including in professional learning communities. Because we often get the question "But what about testing?"—by which teachers generally mean summative assessment—in the introduction we address the link between formative and summative assessment.

Summative assessment is a pervasive topic that includes everything from statewide accountability tests, to local assessments and district benchmark tests, to everyday classroom tests. To grapple with what seems to be an overuse of graded quizzes and testing, educators need to change their view of assessment to one that is about information. The more information we have about students, from both summative and formative assessments, the clearer the picture we have about student learning.

It is important to remember that these probes are used to elicit students' ideas, engage students in discussion about their ideas, and monitor how students' conceptions are changing throughout instruction. They are not intended to be graded. Once you pass judgment on the student with a grade, research shows that the student's learning often shuts down (Black et al. 2003). Use these probes to gather information about your students and motivate them to open up and share their ideas. Grading a formative assessment often does the opposite. When low achievers get back their papers with low scores, the message is, "You are not good enough," and their desire to learn fades. Likewise a good grade also shuts down students' thinking. As long as students see a passing grade, they often ignore the teachers' comments that may indicate ways to improve their work or challenge them to think further. If we are going to use formative assessment effectively, and distinguish it from summative assessment, it is important to get over the pervasive habit of grading every piece of work. Although there are times when it is important to grade work, the probes are not intended for that purpose.

Assessments fall into three different types: formative, summative, and diagnostic. When data are used by teachers to make decisions about next steps for a student or group of students, to plan instruction, and to improve their own practice, they help inform as well as form

Introduction

practice; this is formative assessment. When data are collected at certain planned intervals, and are used to show what students have achieved to date, they provide a summary of progress and are summative assessment (Carlson, Humphrey, and Reinhardt 2003, p. 4).

But what about the third type, diagnostic assessment? Diagnostic assessment is used to uncover a misconception or learning difficulty. When used just for this purpose, it is diagnostic. However, diagnostic assessment becomes formative when the information revealed by the assessment is applied to a situation. Another way to look at the three types and their purposes is to use an analogy.

Have you ever watched the Fox television medical drama *House*? Dr. House is a curmudgeonly medical genius who works with an accomplished group of medical diagnosticians who assess a variety of mysterious illnesses in a teaching hospital. What Dr. House does is very similar to what a teacher does when using the probes. Dr. House and his team collect a variety of data to diagnose an illness a patient has. They often use medical probes to look into the body and see things they would not ordinarily be able to see. The data reveal to Dr. House and his team the cause of the illness. Similarly, the teachers' use of the probes uncovers a variety of problems that may not be obvious to other practitioners who do not have the right tools or deep understandings about student learning.

Naturally, Dr. House and his team do not stop with the diagnosis. They want the patient to get better, so they prescribe the best course of treatment that will take care of the medical problem. The treatment is informed by the diagnosis and any additional data on the patient's condition. Dr. House and his team closely monitor the patient for improvement. In educational assessment, this would be the formative assessment—moving beyond the diagnosis to inform the instructional strategies and direction a teacher will use to help students with their learning problems and achieve conceptual understanding.

At the end of a course of treatment, the patient often comes back for a follow-up checkup to see if he or she is cured. This usually happens after the treatment has ended. In education, summative assessment is usually given after a sequence of instruction or at the end of a course to find out how well a student has learned.

Of course, science assessment is not a life or death decision-making process, and today's teachers would not communicate with their students in the way Dr. House abruptly confronts his patients. Teachers' assessments, however, do involve carefully made choices that have a potentially huge impact on student learning. Assessment is not about just testing anymore, and practicing for test taking may not lead to constructive changes in the classroom. Relying on external test results doesn't help much to improve learning in the immediate sense. Traditionally, summative assessments have not helped teachers adjust instruction for individual students because it takes too long for the data to be returned to schools. Formative assessment, such as the probes in this book, can be used on a regular basis to monitor student progress and modify instruction when it is needed.

Teachers and schools are now looking at classroom assessment more closely and at how to best improve teaching and learning so that the assessments we give at the end of a unit or at the end of a year are better matched to what students have been learning, not what they memorized for a test. By linking diagnostic and formative assessments, such as the probes in this book, to the standards being assessed on summative assessments, teachers get a better picture of what they need to do to move students toward the development of conceptual understanding so that students can be successful on summative assessments.

As depicted in the vignette at the beginning of this introduction, some schools are beginning to shift away from a rigid accountability system and move toward the use of more assessments, such as the ones in this book, that look at the whole picture of student learning. When teachers and districts begin to align these formative assessments with the standards assessed on summative assessments, the student data from these probes can provide powerful predictors of readiness for summative assessment. When teachers design their summative assessments with the standards in mind and a picture of what success is, they can use the assessment probes prior to and throughout instruction to monitor how well students are moving toward the learning targets—and adjust their teaching accordingly. As students become more metacognitive about what they know, think they know, and do not know, they take more responsibility for their learning, which eventually improves their performance on measures of achievement.

As a noted superintendent once succinctly said, "Schools are data rich and information poor" (DuFour, Dufour, and Eaker 2005 p. 40). We hope that the probes in this book will uncover information about your students' thinking that will provide a gold mine of useful data for your classroom and for the improvement of your school's science program. Forging a stronger link between formative and summative assessment will create a better assessment balance and lead to your desired results. Are the desired results improved test scores? They may be the "trees" you are looking at, but the real aim is to look through the trees to see the whole forest. In that forest is the "big picture" of learning that will produce science-literate adults. As you use these probes, remember to look beyond the trees and into the forest!

References

Black, P., C. Harrison, C. Lee, B. Marshall, and D. Wiliam. 2003. *Assessment for learning.* Berkshire, UK: Open University Press.

Carlson, M., G. Humphrey, and K. Reinhardt. 2003. *Weaving science inquiry and continuous assessment.* Thousand Oaks, CA: Corwin Press.

DuFour, R., R. Eaker, and R. Dufour, eds. 2005. *On common ground: The power of professional learning communities.* Bloomington, IN: National Education Service.

Mundry, S., P. Keeley, and C. Landel. 2009. *A leader's guide to science curriculum topic study: Designs, tools, and resources for professional learning.* Thousand Oaks, CA: Corwin Press.

Wiliam, D. 2008. Improving learning in science with formative assessment. In *Assessing science learning,* eds. J. Coffey, R. Douglas, and C. Stearns. Arlington, VA: NSTA Press.

Physical Science and Unifying Themes Assessment Probes

Core Science Concepts	Physical Science									Unifying Themes	
	Sugar Water	Iron Bar	Burning Paper	Nails in a Jar	Salt Crystals	Ice Water	Warming Water	Standing on One Foot	Magnets in Water	Is It a Model?	Is It a System?
Atoms		✓			✓						
Chemical change			✓	✓							
Closed system			✓	✓							
Combustion			✓								
Conservation of matter			✓	✓							
Crystal					✓						
Crystalline lattice					✓						
Dissolving	✓										
Energy						✓	✓				
Force								✓			
Gravity								✓			
Heat							✓				
Ionic bond					✓						
Magnetism									✓		
Mixture	✓										
Model										✓	
Oxidation				✓							
Phase change						✓					
Phases of matter						✓					
Physical change	✓										
Pressure								✓			
System											✓
Temperature						✓	✓				
Thermal energy							✓				
Thermal expansion		✓									
Transfer of energy						✓	✓				
Weight								✓			

Sugar Water

Deanna stirred a teaspoon of sugar into a glass of warm water. The sugar completely dissolved in the water. Put an X next to the statements that are true about the dissolved sugar.

_____ **A** The sugar melts.

_____ **B** The sugar loses mass.

_____ **C** The sugar turns into water molecules.

_____ **D** The sugar forms a mixture with the water.

_____ **E** The sugar can be separated from the water.

_____ **F** The sugar disappears and no longer exists.

_____ **G** The sugar molecules are spread among the water molecules.

_____ **H** The sugar breaks down into the individual atoms that make up sugar.

_____ **I** The sugar chemically combines with the water to form a new substance.

Explain your thinking. Describe what happens to sugar when it dissolves in water.

_____

_____

_____

_____

_____

_____

_____

_____

Sugar Water

Teacher Notes

Purpose

The purpose of this assessment probe is to elicit students' ideas about dissolving. The probe is designed to find out what students think happens to sugar when it dissolves in water.

Related Concepts

dissolving, mixture, physical change

Explanation

The best answers are D, E, and G. A grain of sugar is actually a large collection of sugar molecules. When these grains of sugar are added to the water, they dissolve—forming a mixture called a solution. This solution is formed because water is a polar molecule, in which one part of the molecule has a slight positive charge and the other part has a slight negative charge. Opposite charges attract. When grains of sugar (tiny sugar crystals) are added to water, the positive part of the polar water molecules attracts (through dipole forces) the specific groupings on the sugar molecules (called hydroxyl groups) that have a slight negative charge. This dipole force does not break the molecular bonds in the individual sugar molecules, resulting in individual atoms that make up the sugar. Instead, the force overcomes the intermolecular attraction that holds the large number of individual sugar molecules together in the form of a "sugar grain," or crystal. The sugar molecules become surrounded by the attracted water molecules so that the individual sugar molecules are no longer part of the crystal. This process is repeated until either (a) all the sugar (the solute) is dissolved in the water (the solvent) to form a solution or (b) there are no longer any "unattached" water molecules and no more sugar can dissolve.

When sugar dissolves, it is physically, not chemically, combined with water, and therefore it does not form a new compound. Instead, it forms a mixture—sugar water. The sugar can be separated from the water in solution through evaporation or boiling off of the water. Sugar crystals form again through intermolecular forces among the sugar molecules that are no longer surrounded by the polar water molecules. During the dissolving process, the sugar still exists as a molecular compound. It has not melted because melting involves a change of state and does not require the interaction between two substances as dissolving does. According to the conservation of mass principle, the weight or mass of the sugar remains the same even though it cannot be seen in the solution. It is still there. No additional sugar molecules have been added or taken away after dissolving.

Curricular and Instructional Considerations

Elementary Students

In the elementary grades, students explore a variety of observable physical changes, including dissolving and melting. Mixing sugar in water and evaporating water to recover sugar crystals is a common experience for elementary students. They see the sugar "disappear" but may not understand where it goes. By evaporating the water, they see that it is does "come out" of the solution. This probe is useful in eliciting early ideas about what happens to substances that dissolve. However, a molecular explanation of what happens during dissolving exceeds this grade level.

Middle School Students

In the middle grades, students begin to use particulate ideas to explain phenomena such as sugar dissolving or substances melting. They should be able to differentiate between chemical changes of the same substance (e.g., burning sugar) versus a physical change (e.g., dissolving sugar). Mixtures and solutions are commonly investigated by students at this grade level. However, students at this level may still confuse dissolving with melting, particularly when a liquid is involved as part of the system. At this level the difference between dissolving and melting can be explained by interactions. Middle school students may begin to use the idea of molecules or simple particulate models to explain what happens when the sugar dissolves, although an explanation of the nature of attraction between sugar and water molecules should wait until high school. At this level, students should also be able to use the idea of atoms and molecules to explain how the mass (or weight) is conserved during the dissolving process.

High School Students

Students at the high school level develop a more sophisticated understanding of a particulate model of matter that can be used to explain dissolving. They encounter formal concepts and ideas in chemistry dealing with the attraction among and between particles and their arrangements and begin to develop

an understanding of the hydrogen bonds and the attraction between molecules of a solute and solvent. This more sophisticated particulate model can be developed to help overcome ideas about dissolving and melting being the same process. This probe is useful in finding out if students have changed previously held ideas or if they still hold on to their preconceptions about dissolving, even after formal instruction.

Administering the Probe

Be sure students are familiar with the phenomenon described. It may be useful to demonstrate by dissolving a teaspoon of sugar into a glass of warm water so that they can see that the sugar is no longer visible in the glass of water. For younger students, consider eliminating choices C, G, and H if they have not yet had an opportunity to develop ideas about the structure of matter at the molecular level.

Related Ideas in *National Science Education Standards* (NRC 1996)

K–4 Properties of Objects and Materials

- Objects have many observable properties, including size, weight, shape, color, temperature, and the ability to react with other substances.

5–8 Properties and Changes of Properties in Matter

★ A substance has characteristic properties, such as solubility. A mixture of substances often can be separated into the original substances using one or more of the characteristic properties.

9–12 Structure and Properties of Matter

★ The physical properties of compounds reflect the nature of the interactions among its molecules. These interactions are determined by the structure of the molecule, including the constituent atoms and the distances and angles between them.

Related Ideas in *Benchmarks for Science Literacy* (AAAS 1993 and 2008)

Note: Benchmarks revised in 2008 are indicated by (R). New benchmarks added in 2008 are indicated by (N).

3–5 Structure of Matter

★ Materials may be composed of parts that are too small to be seen without magnification.

- When a new material is made by combining two or more materials, it has properties that are different from the original materials. (R)

6–8 Structure of Matter

- All matter is made up of atoms, which are far too small to be seen directly through a microscope.

★ No matter how substances within a closed system interact with one another, or how they combine or break apart, the total mass of the system remains the same. (R)

★ Indicates a strong match between the ideas elicited by the probe and a national standard's learning goal.

9–12 Structure of Matter

★ Atoms often join with one another in various combinations in distinct molecules or in repeating three-dimensional crystal patterns. An enormous variety of biological, chemical, and physical phenomena can be explained by changes in the arrangement and motion of atoms and molecules.

Related Research

- From an early age through to adulthood, conceptions about dissolving include the following: the solute "disappears," "melts away," "dissolves away," or "turns into water." Older students imagine that as sugar dissolves, it "goes into tiny little bits," "sugar molecules fill spaces between water molecules," or sugar "mixes with water molecules" (Driver et al. 1994).

- Students' ideas about what happens to sugar as it dissolves frequently fail to include the conservation of mass. The gap between the proportion of students who conserved substance but not mass widened between ages 9 and 11 but narrowed in later school years. After age 12, many, but not all, students begin to develop a conception of weight and mass and begin to conserve mass of the solute (Driver et al. 1994).

- Students' ideas about solutions include thinking that sugar solutions are not a single phase but rather that invisible gross particles of sugar are suspended in the solution. They may suggest that the particles can be filtered out or will settle out from the solution. Others see the solute and solvent as a single substance rather than as a homogeneous mixture (Driver et al. 1994).

- Cosgrove and Osborne (1980) sampled children ages 8–17 and found they regarded melting and dissolving as similar processes because they were both gradual.

Suggestions for Instruction and Assessment

- This probe can be used to launch an inquiry-based investigation related to what happens to the mass of sugar during the physical process of dissolving in water. Ask students to predict and explain what will happen to the combined mass of the sugar and water before and after dissolving, then test it.

- Challenge students to come up with a way to separate sugar once it is dissolved in water.

- Ask older students to draw "particulate pictures" to show and explain what happens to the sugar and where it goes when it dissolves in water.

- If students experience only colorless solutions, such as sugar in water, the idea that the solute "disappears" may be reinforced. Provide colored solutes such as drink crystals and use the presence of uniform color throughout the solution to help students understand that this is additional evidence for the matter still existing, even though its form has changed. However, it may still be necessary to provide additional experiences to confirm that the total mass (or weight) does not change.

★ Indicates a strong match between the ideas elicited by the probe and a national standard's learning goal.

- When teaching about physical changes, have students come up with a "rule" that can be used to identify a type of physical change. For example, help students recognize that dissolving involves two materials whereas a phase change such as melting involves one. This can explicitly be used to help students develop a rule to decide if something dissolves or melts.

- With middle school students, use the "Scientists' Idea" strategy to compare their initial ideas about the probe and dissolving with how a chemist would answer the probe. You can use the explanation on pages 12–13 as the chemist's ideas or provide a reading in their text or other resources that describe dissolving. Several other formative assessment strategies described in *Science Formative Assessment: 75 Practical Strategies for Linking Assessment, Instruction, and Learning* (Keeley 2008) can be used with this probe.

Related NSTA Science Store Publications, NSTA Journal Articles, NSTA SciGuides, NSTA SciPacks, and NSTA Science Objects

American Association for the Advancement of Science (AAAS). 1993. *Benchmarks for science literacy.* New York: Oxford University Press.

Ashbrook, P. 2006. Mixing and making changes. *Science & Children* (Feb.): 29–31.

Kessler, J., and P. Galvan. 2006. Dynamics of dissolving. *Science & Children* (Feb.): 45–46.

National Science Teachers Association (NSTA). 2005. Properties of objects and materials. NSTA SciGuide. Online at *http://learningcenter.nsta.org/ product_detail.aspx?id=10.2505/5/SG-01.*

Related Curriculum Topic Study Guides
(Keeley 2005)
"Mixtures and Solutions"
"Physical Properties and Change"

References

American Association for the Advancement of Science (AAAS). 1993. *Benchmarks for science literacy.* New York: Oxford University Press.

American Association for the Advancement of Science (AAAS). 2008. Benchmarks for science literacy online. *www.project2061.org/publications/ bsl/online*

Cosgrove, M., and R. Osborne. 1980. Physical change. LISP Working Paper 26. Hamilton, New Zealand: University of Waikato, Science Education Research Unit.

Driver, R., A. Squires, P. Rushworth, and V. Wood-Robinson. 1994. *Making sense of secondary science: Research into children's ideas.* London: RoutledgeFalmer.

Keeley, P. 2005. *Science curriculum topic study: Bridging the gap between standards and practice.* Thousand Oaks, CA: Corwin Press.

Keeley, P. 2008. *Science formative assessment: 75 practical strategies for linking assessment, instruction, and learning.* Thousand Oaks, CA: Corwin Press.

National Research Council (NRC). 1996. *National science education standards.* Washington, DC: National Academy Press.

Iron Bar

Nate measured an iron bar. He put the iron bar in the hot sun. When he measured the bar after it had been in the sun, it was slightly longer. Which sentence best describes what happened to the iron atoms after the bar was left in the hot sun?

A   The number of atoms increased.

B   The size of the atoms increased.

C   The space between each atom increased.

D   The air in the spaces between the iron atoms expanded.

E   Some of the atoms began to melt and spread out further in the bar.

F   The heat caused the atoms to flow around the bar and pushed it outward.

Explain your thinking about what happens to atoms when a metal is heated. You may draw pictures to support your explanation.

_____

_____

_____

_____

_____

_____

_____

_____

_____

Iron Bar

Teacher Notes

Purpose

The purpose of this assessment probe is to elicit students' ideas about atoms. The probe is specifically designed to determine whether students can use the idea of atoms to explain why a metal expands when heated. Furthermore, older students' explanations may reveal whether these students use the kinetic molecular theory to explain how heating causes the atoms to vibrate more, thus pushing the atoms apart.

Related Concepts

atoms, thermal expansion

Explanation

The best answer is C: The space between each atom increased. Thermal expansion of metals involves the tendency of a metal to increase in volume in response to an increase in temperature. As a metal object is heated, its atoms vibrate in place more vigorously and, as a result, increase the separation between individual atoms. This slight increase in the empty space between atoms results in a cumulative change in the measurable volume of the object. The object expands. When the object expands, no new atoms are added. Although the length or width of the metal object may increase, the size of the atoms it is made of stays the same. It is the space between the atoms that increases and contributes to an increase in volume. As the atoms of a solid gain energy, they vibrate more in place. Unlike a gas, which is free to move about, the metal atoms maintain their general positions.

Curricular and Instructional Considerations

Elementary Students

In the elementary grades, students observe macroscopic properties of matter, including changes caused by heating. Their observations focus on objects and materials. Explanations of phenomena that use the idea of atoms should wait until middle school or when students are ready to use this abstract idea.

Middle School Students

In the middle grades, students begin to use atomic and molecular ideas to explain phenomena. They begin to relate the expansion and contraction associated with the heating and cooling of substances to the position and motion of particles. However, students at this level may still confuse the properties of the material or substance with the properties of the atoms or molecules of which they are made.

High School Students

Students at the high school level should be able to use ideas about atomic/molecular motion to explain phenomena from a microscopic view. They should be able to distinguish between the observable properties of a substance and the properties of the atoms making up the substance. However, many students at this stage will attribute expansion of the solid material to an increase in the amount of matter and/or an increase in the size of the atoms rather than to the space between them.

Administering the Probe

Make sure students understand the phenomenon. Consider providing real-life examples of similar phenomena, such as the space between the metal joints on a bridge expanding in the summer or a metal door that sometimes scrapes against the floor on a hot summer day. The ball and ring apparatus sold in science supply stores can also be used by demonstrating how the metal ball can no longer pass through the ring when it has been heated.

Related Ideas in *National Science Education Standards* (NRC 1996)

K–4 Properties of Objects and Materials

- Objects have many observable properties, including size, weight, shape, color, temperature, and the ability to react with other substances.
- Objects are made of one or more materials, such as paper, wood, or metal. Objects can be described by the properties of the materials from which they are made.

5–8 Transfer of Energy

- Energy is transferred in many ways.

9–12 Conservation of Energy and the Increase in Disorder

- ★ Heat consists of random motion and the vibrations of atoms, molecules, and ions. The higher the temperature, the greater the atomic or molecular motion.

★ Indicates a strong match between the ideas elicited by the probe and a national standard's learning goal.

Related Ideas in *Benchmarks for Science Literacy* (AAAS 1993 and 2008)

Note: Benchmarks revised in 2008 are indicated by (R). New benchmarks added in 2008 are indicated by (N).

K–2 Structure of Matter

- Objects can be described in terms of their properties. Some properties, such as hardness and flexibility, depend on what material the object is made of, and some properties, such as size and shape, do not. (R)

- Things can be done to materials to change some of their properties, but not all materials respond the same way to what is done to them.

3–5 Structure of Matter

- Heating and cooling can cause changes in the properties of materials, but not all materials respond the same way to being heated and cooled. (R)

- Materials may be composed of parts that are too small to be seen without magnification.

- All materials have certain physical properties, such as strength, hardness, flexibility, durability, resistance to water and fire, and ease of conducting heat. (N)

6–8 Structure of Matter

- All matter is made up of atoms that are far too small to be seen directly through a microscope.

- ★ Atoms and molecules are perpetually in motion. Increased temperature means greater average energy of motion, so most substances expand when heated. In solids, the atoms are closely locked in position and can only vibrate.

6–8 Energy Transformations

- Thermal energy is transferred through a material by the collision of atoms within the material. Over time, the thermal energy tends to spread out through a material and from one material to another if they are in contact. (R)

9–12 Structure of Matter

- ★ An enormous variety of biological, chemical, and physical phenomena can be explained by changes in the arrangement and motion of atoms and molecules.

Related Research

- Students of all ages show a wide range of beliefs about the nature and behavior of particles. For example, they attribute macroscopic properties to particles; do not accept the idea there is empty space between particles; and have difficulty accepting the intrinsic motion of solids, liquids, and gases (AAAS 1993).

- Children frequently consider atoms of a solid to have all or most of the macro-properties they associate with the solid (Driver et al. 1994).

- Some older students (ages 11–16) who had learned about the kinetic molecular theory of matter attempted to explain conductivity phenomena in terms of particulate

---

★ Indicates a strong match between the ideas elicited by the probe and a national standard's learning goal.

ideas. However, these ideas were not used spontaneously by most of the students interviewed. When they did use particle ideas, they had a tendency to attribute expanding (getting bigger) to the properties of the atoms (Driver et al. 1994).

- Children's naive view of particulate matter is based on a "seeing is believing" principle in which they tend to use sensory reasoning. Being able to accommodate a scientific particle model involves overcoming cognitive difficulties of both a conceptual and perceptive nature (Kind 2004).

Suggestions for Instruction and Assessment

- Demonstrate the phenomenon with the ball and ring apparatus sold by most science supply companies. Before heating, the ball easily passes through the ring. After the ball is heated, it expands and will no longer pass through the ring when warm. Ask students to describe what happened to the ball itself. Then ask students to describe what happened to the atoms that make up the ball. Ask them to describe the difference between the phenomenon at the substance level versus the atomic level.

- Encourage students to draw "atomic pictures" of what they think is happening to the atoms as the metal is heated.

- Have students generate a list of other things that expand when heated and describe what is happening to the atoms or molecules.

- Ask students what happens to the metal when it cools down. Have them explain what happens at both the substance and atomic level.

- The PRISMS (Phenomena and Representations for Instruction of Science in Middle Schools) website at *http://prisms. mmsa.org* has a collection of web-based phenomena and representations aligned to the *Benchmarks for Science Literacy* (AAAS 1993) and Curriculum Topic Study Guides (Keeley 2005) that can be used to help students understand why metals expand when heated.

Related NSTA Science Store Publications, NSTA Journal Articles, NSTA SciGuides, NSTA SciPacks, and NSTA Science Objects

Association for the Advancement of Science (AAAS). 2001. *Atlas of science literacy.* Vol. 1. (See "Atoms and Molecules" map, pp. 54–55.) Washington, DC: AAAS.

Michaels, S., A. Shouse, and H. Schweingruber. 2008. *Ready, set, science! Putting research to work in K–8 science classrooms.* Washington, DC: National Academies Press.

Robertson, W. 2007. *Chemistry basics: Stop faking it! Finally understanding science so you can teach it.* Arlington, VA: NSTA Press.

Related Curriculum Topic Study Guides

(Keeley 2005)
"Particulate Nature of Matter (Atoms and Molecules)"
"Solids"

References

American Association for the Advancement of Science (AAAS). 1993. *Benchmarks for science literacy.* New York: Oxford University Press.

American Association for the Advancement of Science (AAAS). 2008. Benchmarks for science literacy online. *www.project2061.org/publications/bsl/online*

Driver, R., A. Squires, P. Rushworth, and V. Wood-Robinson. 1994. *Making sense of secondary science: Research into children's ideas.* London: RoutledgeFalmer.

Keeley, P. 2005. *Science curriculum topic study: Bridging the gap between standards and practice.* Thousand Oaks, CA: Corwin Press.

Kind, V. 2004. *Beyond appearances: Students' misconceptions about basic chemical ideas.* 2nd ed. Durham, England: Durham University School of Education. Also available at *www.rsc.org/education/teachers/learnnet/miscon.htm*

National Research Council (NRC). 1996. *National science education standards.* Washington, DC: National Academy Press.

Burning Paper

Carey crumpled a wad of paper and placed it in a large glass jar. He recorded the total mass of the jar (including the air in the jar), the paper, the lid, and a match.

Carey lit the match, quickly put it in the jar, and sealed the lid. Most of the paper burned. He saw smoke in the jar and black ashes left from the paper.

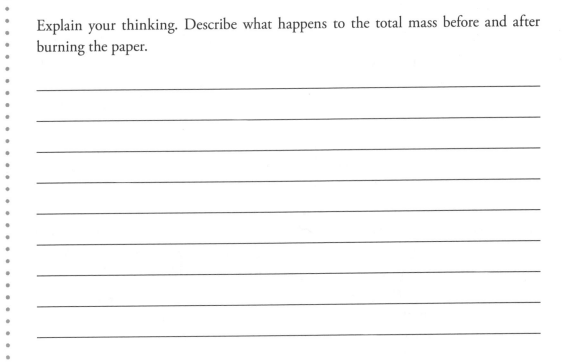

Which sentence best describes the total mass of the jar, lid, paper, and match before burning compared with the total mass after burning?

A The total mass after burning is greater.

B The total mass after burning is less.

C The total mass before and after burning is the same.

Explain your thinking. Describe what happens to the total mass before and after burning the paper.

_____

_____

_____

_____

_____

_____

_____

_____

Burning Paper

Teacher Notes

Purpose

The purpose of this assessment probe is to elicit students' ideas about conservation of matter during combustion. The probe is designed to find out if students think the mass changes as paper burns inside a closed system.

Related Concepts

chemical change, closed system, combustion, conservation of matter

Explanation

The best answer is C: The total mass before and after burning is the same. Burning is an example of combustion—a chemical change in which a substance containing hydrocarbons combines with oxygen to produce carbon dioxide and water. It also releases energy in the form of heat and light. When the carbon and hydrogen of the hydrocarbon-containing substance (i.e., the paper) chemically combine with the oxygen, the remaining materials may appear as ash, the solid remains of a fire. Although the hydrocarbons appear to "vanish" during the reaction with oxygen to form gaseous carbon dioxide and water vapor, the total mass or weight of the reactants (hydrocarbon-containing substance and oxygen in the air) and products (carbon dioxide, water, and ash) remain the same. In a closed system containing air, a piece of paper, and a match, no mass or weight is added or lost as the paper burns.

Gases play a big part in the interaction that occurs inside the jar. Many students have observed wood burning in a fireplace or other structure and they see that many pounds of wood seem to "disappear" with only ash left. What they do not see are the many pounds of

gas given off that leave through the chimney. It is important for students to think about the interaction of all materials inside the jar.

Curricular and Instructional Considerations

Elementary Students

In the elementary grades, students begin developing ideas about changes in objects and materials. They can recognize the formation of soot or ash as a change in the appearance of the paper. Upper-elementary students begin to distinguish between physical and chemical changes on the basis of changes in observable properties. Conservation of matter in the elementary grades focuses on parts and wholes of objects and changes of state. Although the chemical details are too sophisticated to be addressed at this age level, the probe can be used to find out elementary students' intuitive ideas about the conservation of matter in a closed system.

Middle School Students

In the middle grades, students link ideas about chemical change with formation of new substances. Burning (combustion) is commonly used as an example of a chemical change that results in a new substance with properties that differ from the original substance. These basic ideas about chemical change are included as grade-level expectations in the national standards. However, the mechanism of that change, explained by the interaction among hydrogen, carbon, and oxygen atoms,

is a more sophisticated idea developed in high school. The probe is useful in determining students' initial ideas about what burning and the combustion process are.

By the end of middle school, all students should know that matter or mass is conserved in a closed system as well as in chemical reactions. Conservation ideas about objects begin in elementary grades and increase in cognitive sophistication as the ideas of atoms, interactions, transformations, and closed systems are considered. Transformation of matter is addressed in middle school, although it remains a difficult concept and one in which students may have difficulty applying conservation reasoning. The notion that gases are involved in the interaction may be missing, and the "disappearance" of the paper may influence students' thinking that the paper is breaking down and losing mass. Knowing the ideas that students hold prior to learning that oxygen combines with substances in the paper during a combustion reaction is useful in designing learning experiences that challenge their intuitive notions influenced by observation.

High School Students

Students at the high school level make a transition from a basic understanding of types of chemical changes, including composition, decomposition, and single and double replacement reactions, to understanding the mechanism for the reaction. Conservation of matter or mass at the high school level is an idea applied to other matter-related ideas in biological, physical, and geological contexts.

The probe is useful in determining whether students recognize a closed system as justification for matter or mass being conserved during a chemical change. The probe is also useful in determining whether students still hold on to preconceived ideas about burning, even after they have received middle school instruction targeted toward the idea that in a combustion reaction, oxygen combines with certain materials to form carbon dioxide and water.

Administering the Probe

Be sure students understand that the air, paper, and match are contained in a sealed jar and nothing can enter or escape from the jar. It may help to have visual props for this probe. Light a match and seal it in a jar containing a crumpled wad of paper. Have students observe the paper as it burns. Ask students to consider what happened to the total weight or mass of the system. Note: You may wish to substitute the word *mass* with the word *weight* if using this probe with elementary school students.

The probe "Nails in a Jar" (p. 31), along with several probes in Volume 1 of this series (Keeley, Eberle, and Farrin 2005), can be used to further probe students' ideas about conservation of matter or chemical changes involving oxygen.

Related Ideas in *National Science Education Standards* (NRC 1996)

K–4 Properties of Objects and Materials

- Objects have many observable properties, including size, weight, shape, color, temperature, and the ability to react with other substances.

5–8 Properties and Changes in Properties of Matter

★ Substances react chemically in characteristic ways with other substances to form new substances (compounds) with different characteristic properties. In chemical reactions, the total mass is conserved.

9–12 Structure of Atoms

- Matter is made up of minute particles called atoms.

9–12 Chemical Reactions

- Chemical reactions occur all around us.

Related Ideas in *Benchmarks for Science Literacy* (AAAS 1993 and 2008)

Note: Benchmarks revised in 2008 are indicated by (R). New benchmarks added in 2008 are indicated by (N).

K–2 Structure of Matter

- Objects can be described in terms of the materials they are made of (e.g., clay, cloth,

---

★ Indicates a strong match between the ideas elicited by the probe and a national standard's learning goal.

paper) and their physical properties (e.g., color, size, shape, weight, texture, flexibility).

• Things can be done to materials to change some of their properties, but not all materials respond the same way to what is done to them.

3–5 Structure of Matter

• When a new material is made by combining two or more materials, it has properties that are different from the original materials.

★ No matter how parts of an object are assembled, the weight of the whole object made is always the same as the sum of the parts, and when a thing is broken into parts, the parts have the same total weight as the original object. (R)

6–8 Structure of Matter

★ An especially important kind of reaction among substances involves the combination of oxygen with something else, as in burning or rusting.

★ No matter how substances within a closed system interact with one another, or how they combine or break apart, the total mass of the system remains the same. (R)

• Substances react chemically in characteristic ways with other substances to form new substances with different characteristic properties.

★ The idea of atoms explains chemical reactions: When substances interact to form new substances, the atoms that make up the molecules of the original substances combine in new ways. (N)

9–12 Structure of Matter

• Atoms often join with one another in various combinations in distinct molecules or in repeating three-dimensional crystal patterns. An enormous variety of biological, chemical, and physical phenomena can be explained by changes in the arrangement and motion of atoms and molecules.

Related Research

• Studies of 11- and 12-year-olds' ideas about the role of air in burning suggest that most know that air is needed for burning, but the function of air is not generally understood (Driver et al. 1994).

• Students may realize that oxygen is necessary for combustion but may not understand how it interacts with the material. Some combustibles are said to have "melted" or "evaporated," or the combustible substance is thought to be made up of the substances that eventually appear as products (Driver et al. 1994).

• More than half of a group of 15-year-olds considered to have "above average ability" predicted loss of mass on the combustion of a sample of iron wool (Driver et al. 1994).

• Many students do not recognize the quantitative aspects of a chemical change and the conservation of overall mass (Driver et al. 1994).

• Middle and high school students' thinking about chemical change tends to be dominated by the obvious features of the change. Some students think that when something is burned in a closed container,

---

★ Indicates a strong match between the ideas elicited by the probe and a national standard's learning goal.

it will weigh more because they see the smoke that was produced (AAAS 1993).

- For chemical reactions that evolve gas, mass conservation is more difficult for students to grasp (AAAS 1993). If a chemical reaction results in the apparent disappearance of some materials, students may not know that mass is conserved (Driver et al. 1994).

Suggestions for Instruction and Assessment

- This probe can be followed up as an inquiry-based demonstration. Ask the question and encourage students to commit to a prediction. Test it by finding the total mass of a jar, lid, paper, and match and then burning the paper in the jar with the match inside and the lid tightly sealed. Have students discuss the evidence and connect the results to the scientific principle of conservation of mass during a chemical reaction: Mass is not created or destroyed in a chemical reaction, but atoms and molecules are rearranged to form new products. However, be sure students are finding the mass within the limits of precision of the scale that is used.

- Reinforce the idea of conservation of mass during a chemical reaction in which a gas is a reactant by exploring additional reactions in closed systems. For example, clean a Ping-Pong-ball-size piece of iron wool (commonly called "steel wool") by dipping it in vinegar and then drying it.

Place it in a flask and stretch a deflated balloon over the top of the flask. Record the mass of this system before and after the wool "rusts" and the balloon gets "sucked in" to the flask. Discuss the observations and connect the results to the scientific principle of conservation of mass during a chemical reaction, placing particular emphasis on the phenomenon of "disappearing gas."

- Reinforce the idea of conservation of mass during a chemical reaction in which a gas is produced by exploring reactions of everyday substances in closed systems. For example, record the mass of an effervescent tablet in a balloon that is sealed over a flask of water before and after the tablet is dropped into the water. Discuss the observations and connect the results to the scientific principle of conservation of mass during a chemical reaction, placing particular emphasis on the evidence of gas production.

- Help students draw parallels between the types of chemical change that involve combination with oxygen, such as oxidation and combustion reactions.

- With older students, connect this probe to the history of science by sharing how Antoine Lavoisier's idea of conservation of matter became the centerpiece of the modern science of chemistry. Recount Lavoisier's careful measurement of substances involved in burning to show that there was no net gain or loss of weight.

Related NSTA Science Store Publications, NSTA Journal Articles, NSTA SciGuides, NSTA SciPacks, and NSTA Science Objects

American Association for the Advancement of Science (AAAS). 2007. *Atlas of science literacy.* Vol. 2. (See "The Chemical Revolution" map, pp. 80–81.) Washington, DC: AAAS

Cobb, C., and M. L. Fetterolf. 2005. *The joy of chemistry: The amazing science of familiar things.* Amherst, NY: Prometheus Books.

Keeley, P. 2005. *Science curriculum topic study: Bridging the gap between standards and practice.* Thousand Oaks, CA: Corwin Press.

Keeley, P., F. Eberle, and L. Farrin. 2005. *Uncovering student ideas in science: 25 formative assessment probes.* Arlington, VA: NSTA Press.

National Science Teachers Association (NSTA). 2005. Properties of objects and materials. NSTA SciGuide. Online at *http://learningcenter.nsta.org/product_detail.aspy?id=10.2505/5/SG-01.*

Robertson, W. 2007. *Chemistry basics: Stop faking it! Finally understanding science so you can teach it.* Arlington VA: NSTA Press.

Related Curriculum Topic Study Guides

(Keeley 2005)
"Chemical Properties and Change"
"Conservation of Matter"

References

American Association for the Advancement of Science (AAAS). 1993. *Benchmarks for science literacy.* New York: Oxford University Press.

American Association for the Advancement of Science (AAAS). 2008. Benchmarks for science literacy online. *www.project2061.org/publications/bsl/online*

Driver, R., A. Squires, P. Rushworth, and V. Wood-Robinson. 1994. *Making sense of secondary science: Research into children's ideas.* London: RoutledgeFalmer.

Keeley, P. 2005. *Science curriculum topic study: Bridging the gap between standards and practice.* Thousand Oaks, CA: Corwin Press.

Keeley, P., F. Eberle, and L. Farrin. 2005. *Uncovering student ideas in science: 25 formative assessment probes.* Vol. 1. Arlington, VA: NSTA Press.

National Research Council (NRC). 1996. *National science education standards.* Washington, DC: National Academy Press.

Nails in a Jar

Jake put a handful of wet, iron nails in a glass jar. He tightly closed the lid and set the jar aside. After a few weeks, he noticed that the nails inside the jar were rusty. Which sentence best describes what happened to the total mass of the sealed jar after the nails rusted?

A The mass of the jar and its contents increased.

B The mass of the jar and its contents decreased.

C The mass of the jar and its contents stayed the same.

Select the answer that best matches your thinking. Explain what happened to the mass before and after the nails rusted.

_____

_____

_____

_____

_____

_____

_____

_____

_____

_____

Nails in a Jar

Teacher Notes

Purpose

The purpose of this assessment probe is to elicit students' ideas about conservation of matter during a chemical change (oxidation). The probe is designed to find out if students think the mass changes as rust forms inside a closed system.

Related Concepts

chemical change, closed system, conservation of matter, oxidation

Explanation

The best answer is C: The mass of the jar and its contents stayed the same. Rusting is an example of oxidation—a chemical change in which electrons from the iron atoms are transferred to the oxygen atoms, resulting in the formation of a new compound. The chemical equation for this change is $4Fe + 3O_2 \rightarrow 2Fe_2O_3$. Oxygen chemically combines with iron atoms on the surface of the nail. As a result, additional mass is added to the nail as a new compound (iron oxide) is formed. Although the appearance of rust on the surface of the nail makes it look as if it is "breaking down," it is actually gaining mass as it changes from iron to iron oxide. However, because the source of the oxygen is from the air that was inside the jar, any added mass to the nail surface can be attributed to an equal loss of mass from the air in this closed system. No mass is added or lost in the total system.

Curricular and Instructional Considerations

Elementary Students

In the elementary grades, students begin developing ideas about changes in objects and mate-

rials. They can recognize rust as a change in the appearance of the nail. Upper-elementary students begin distinguishing between physical and chemical changes on the basis of observation of changes in properties. Rust is often used as an example of a chemical change. This probe may be useful in determining students' early notions of rusting, particularly whether they view rusting as a "decomposing" process. Conservation of matter in the elementary grades focuses on parts and wholes of objects and changes of state. Although the chemical change context of this probe is rather sophisticated for elementary students, it can be used to find out elementary students' intuitive ideas about the change from a shiny nail to a rusted nail.

Middle School Students

In the middle grades, students link ideas about chemical change with formation of compounds, including basic ideas about oxidation. Rust is commonly used as an example of a chemical change that results in a new compound with properties that differ from the original substance. These basic ideas about chemical change are included as grade-level expectations in the national standards. However, the mechanism of that change, explained by the interaction between iron and oxygen atoms, is a more sophisticated idea developed in high school. The probe is useful in determining students' initial ideas about what rust and the rusting process are.

By the end of middle school, all students should know that matter or mass is conserved in a closed system as well as in chemical reac-

tions. Conservation ideas about objects begin in elementary grades and increase in cognitive sophistication as the ideas of atoms, interactions, transformations, and closed systems are considered. Transformation of matter is addressed in middle school, although it remains a difficult concept and one to which students may have difficulty applying conservation reasoning. The notion that gases are involved in the transformation may be missing, and the appearance of the rusty nail may influence students' thinking that the nail is breaking down and losing mass. Knowing the ideas students hold prior to learning that oxygen combines with iron during the rusting process is useful in designing learning experiences that challenge their intuitive notions that have been influenced by observation.

High School Students

Students at the high school level make a transition from a basic understanding of types of chemical changes, including oxidation, to understanding the mechanism for oxidation. Conservation of matter at the high school level is implicit in other matter-related ideas in biological, physical, and geological contexts. The probe is useful in determining whether students recognize a closed system as justification for matter being conserved during a chemical change. The probe is also useful in determining whether students still hold on to preconceived ideas about rusting, even after they have received middle school instruction targeted toward the idea that oxygen combines with the iron to form rust.

Administering the Probe

Be sure students understand that the nails are contained in a sealed jar and nothing can enter or escape from the jar. It may help to have visual props for this probe. Seal five clean, wet, iron nails (not galvanized) in a jar. Have students observe the nails. After a few weeks, observe the jar again and consider what happened to the mass.

The probe "Burning Paper" (p. 23) can be used to further probe students' ideas about conservation of matter during a chemical change involving oxygen. "Nails in a Jar" can be combined with "The Rusty Nails" in Volume 1 of this series to further probe ideas about rusting in an open system in which the mass increases (Keeley, Eberle, and Farrin 2005).

Related Ideas in *National Science Education Standards* (NRC 1996)

. . . .

K–4 Properties of Objects and Materials

- Objects have many observable properties, including size, weight, shape, color, temperature, and the ability to react with other substances.

5–8 Properties and Changes in Properties of Matter

★ Substances react chemically in characteristic ways with other substances to form new substances (compounds) with different characteristic properties. In chemical reactions, the total mass is conserved.

9–12 Structure of Atoms

- Matter is made up of minute particles called atoms.

9–12 Chemical Reactions

- Chemical reactions occur all around us.
★ A large number of important reactions involve the transfer of either electrons (oxidation/reduction reactions) or hydrogen ions (acid/base reactions) between reacting ions, molecules, or atoms.

Related Ideas in *Benchmarks for Science Literacy* (AAAS 1993 and 2008)

. . . .

Note: Benchmarks revised in 2008 are indicated by (R). New benchmarks added in 2008 are indicated by (N).

K–2 Structure of Matter

- Objects can be described in terms of the materials they are made of (e.g., clay, cloth, paper) and their physical properties (e.g., color, size, shape, weight, texture, flexibility).
- Things can be done to materials to change some of their properties, but not all materials respond the same way to what is done to them.

3–5 Structure of Matter

- When a new material is made by combining two or more materials, it has properties that are different from the original materials.
- No matter how parts of an object are assembled, the weight of the whole object made is always the same as the sum of the

---

★ Indicates a strong match between the ideas elicited by the probe and a national standard's learning goal.

parts, and when a thing is broken into parts, the parts have the same total weight as the original object. (R)

6–8 Structure of Matter

• Because most elements tend to combine with others, few elements are found in their pure form.

★ An especially important kind of reaction between substances involves the combination of oxygen with something else, as in burning or rusting.

★ No matter how substances within a closed system interact with one another, or how they combine or break apart, the total mass of the system remains the same. The idea of atoms explains the conservation of matter. If the number of atoms stays the same no matter how they are rearranged, then their total mass stays the same.

★ The idea of atoms explains chemical reactions: When substances interact to form new substances, the atoms that make up the molecules of the original substances combine in new ways. (N)

9–12 Structure of Matter

• Atoms often join with one another in various combinations in distinct molecules or in repeating three-dimensional crystal patterns. An enormous variety of biological, chemical, and physical phenomena can be explained by changes in the arrangement and motion of atoms and molecules.

Related Research

• Middle and high school students' ideas about chemical change tend to be dominated by the obvious features of the change (AAAS 1993). For example, students are likely to think that when something rusts it weighs less because it looks as if parts of the metal are being "eaten away" or the powdery rust is less substantive than the iron (Driver et al. 1994).

• Some students describe rust as a type of mold (Driver et al. 1994).

• Some students think that rust comes from the nail itself. Some students explain that it is already under the surface of the nail and is exposed during the rusting process (Driver et al. 1994).

• Students' everyday experiences with rusting often involve iron getting wet. Consequently students are likely to think that rusting happens as a result of the water eating away at the metal, rather than rusting being an interaction with oxygen in the air (Driver et al. 1994).

• In a survey conducted of English 15-year-old students, one-third said the rusty nail would lose mass, one-third said it would gain mass, and one-third said its mass would stay the same. Of these students, just over 10% of those studying chemistry said the mass would increase because the mass of the rust is added to the mass of the nail. There was no indication from their response that the iron in the nail was involved in the formation of rust. Others who understood the reaction explained that the mass would

---

★ Indicates a strong match between the ideas elicited by the probe and a national standard's learning goal.

not change because oxygen doesn't weigh anything (Driver et al. 1994).

- Some students use the "taught" ideas about oxidation but adapt them to their intuitive notions of rusting using reasoning such as "the oxygen dissolves some of the iron" (Driver et al. 1994).

- Studies show that the way students perceive a change may influence their ideas about conservation during that change. For example, if their view is dominated by the appearance of a new material, they may think mass has been added (AAAS 1993).

Suggestions for Instruction and Assessment

- Using nongalvanized iron nails, have students carry out an investigation to test their ideas. Use their observations to resolve the differences between their prediction and results. Note: Rusty nails by themselves do not present a safety hazard; however, if students handle the nails, have them wash their hands afterward.

- Develop an understanding of open versus closed systems and explicitly link conservation of matter during a chemical change to changes within a closed system. Carry out an investigation to compare the change in mass as iron nails rust in a closed versus an open system.

- Help students understand the idea that there is oxygen in the air that is also being transformed in the jar. Explicitly develop the idea that air is a substance and has mass; revisit this idea throughout the grade levels.

- Help students draw parallels between other types of chemical change that involve combination with oxygen, such as combustion reactions, to rusting.

- Extend the idea of conservation of matter in a closed system by changing the context to a physical change (see the probe "Ice Cubes in a Bag" in Volume 1 of this series; Keeley, Eberle, and Farrin 2005) or a chemical change in a living system (see the probe "Seedlings in a Jar" in Volume 1 of this series; Keeley, Eberle, and Farrin 2005).

Related NSTA Science Store Publications, NSTA Journal Articles, NSTA SciGuides, NSTA SciPacks, and NSTA Science Objects

American Association for the Advancement of Science (AAAS). 2001. *Atlas of science literacy.* Vol. 1. (See "Conservation of Matter," pp. 56–57.) Washington, DC: AAAS.

Cobb, C., and M. L. Fetterolf. 2005. *The joy of chemistry: The amazing science of familiar things.* Amherst, NY: Prometheus Books.

Drigel, G. S., A. M. Sarquis, and M. D'Agostino. 2008. Corrosion in the classroom. *The Science Teacher* (Apr./May): 50–56.

National Science Teachers Association (NSTA). 2005. Properties of objects and materials. NSTA SciGuide. Online at *http://learningcenter.nsta.org/product_detail.aspx?id=10.2505/5/SG-01.*

Robertson, W. 2007. *Chemistry basics: Stop faking it! Finally understanding science so you can teach it.* Arlington VA: NSTA Press.

Related Curriculum Topic Study Guides

(Keeley 2005)

"Chemical Properties and Change"

"Conservation of Matter"

References

American Association for the Advancement of Science (AAAS). 1993. *Benchmarks for science literacy.* New York: Oxford University Press.

American Association for the Advancement of Science (AAAS). 2008. Benchmarks for science literacy online. *www.project2061.org/publications/bsl/online*

Driver, R., A. Squires, P. Rushworth, and V. Wood-Robinson. 1994. *Making sense of secondary science: Research into children's ideas.* London: RoutledgeFalmer.

Keeley, P. 2005. *Science curriculum topic study: Bridging the gap between standards and practice.* Thousand Oaks, CA: Corwin Press.

Keeley, P., F. Eberle, and L. Farrin. 2005. *Uncovering student ideas in science: 25 formative assessment probes.* Vol. 1. Arlington, VA: NSTA Press.

National Research Council (NRC). 1996. *National science education standards.* Washington, DC: National Academy Press.

Salt Crystals

Boris and his friends were looking at tiny crystals of salt with a magnifying glass. They wondered what they would see if they had a magnifying device powerful enough to see the atoms. This is what they thought they might see:

Boris: "The atoms would be packed tightly together. They would look like a solid material without any empty spaces between the atoms."

Portia: "I think I would see vibrating atoms arranged in an orderly way with spaces between them. There would be nothing in the spaces, not even air."

Vivi: "I think I would see atoms not moving and arranged in an orderly way with spaces between them."

Leif: "I think I would see vibrating atoms arranged in an orderly way. There would be space between the atoms. The space would be filled with air."

Whit: "I think I would see atoms in the shape of small cubes. Each of these small cubes would join together to form a larger cube of salt."

Guillermo: "I think I would see lots of vibrating atoms connected together by little lines. The lines connecting each atom give them a definite cube shape."

Shantidas: "I think I would see individual atoms moving from place to place. They would be moving all about the inside of the crystal shape."

Which student do you most agree with and why? Draw a picture that best represents how you think you would see the atoms in a salt crystal if you had a powerful magnifying device. Explain your picture.

_____

_____

_____

Salt Crystals

Teacher Notes

Purpose

The purpose of this assessment probe is to elicit students' ideas about crystalline solids. The probe is specifically designed to determine how students think atoms are arranged and move in a crystalline lattice.

Related Concepts

atoms, crystal, crystalline lattice, ionic bond

Explanation

The best answer is Portia's: "I think I would see vibrating atoms arranged in an orderly way with spaces between them. There would be nothing in the spaces, not even air." Salt is an example of a crystalline ionic lattice. A salt crystal is made up of an orderly repeating array of sodium and chloride ions. This repeating array is caused by the electrostatic attrac-

tion between negatively and positively charged atoms called ions and forms the salt crystal's distinct cuboidal shape. The tiny crystals are made up of the atoms (in the form of ions). They are in the form of a solid in which the atoms are closely locked in position and can only vibrate. They are not free to move around as in a gas. There is empty space between the atoms that make up the salt crystal. There is no air in these spaces because the material is salt (sodium chloride), not a mixture of salt and air. The crystalline matter is sodium and chlorine atoms only. Sometimes models, such as ball and stick models, depict sodium chloride (table salt) as a repeating cuboidal three-dimensional array of atoms connected by lines representing the ionic bonds. These lines are not actual physical structures but rather represent the attraction among the ions.

Curricular and Instructional Considerations

Elementary Students

In the elementary grades, students observe macroscopic properties of matter and details they can see using magnifiers. Their observations focus on the features of objects and materials. Using magnifiers, they can see that salt has a cuboidal shape. However, explaining that microscopic structure in terms of atoms exceeds expectations for this grade level.

Middle School Students

In the middle grades, students begin to use atomic and molecular ideas to explain phenomena and structural arrangements. They distinguish between molecular substances and crystalline lattices, although the details of ionic and covalent bonding can wait until high school. They should know that solids are rigid structures made up of atoms and that the atoms, with some empty space between them, can only vibrate in place, not move about.

High School Students

Students at the high school level should be able to use ideas about atomic/molecular motion to explain phenomena and structural arrangement from a microscopic view. They should be able to explain the difference between ionically bonded compounds and other types of chemical bonds. They frequently use ball and stick models to explain structure and behavior. However, even though they may understand what an ionic bond is, they may still hold on to misconceptions about the space between atoms.

Administering the Probe

This probe is most appropriate at the middle and high school levels. Consider having students examine grains of salt macroscopically before answering the probe.

Related Ideas in *National Science Education Standards* (NRC 1996)

K–4 Properties of Objects and Materials

- Objects have many observable properties, including size, weight, shape, color, temperature, and the ability to react with other substances.

9–12 Structure of Atoms

- Matter is made up of minute particles called atoms, and atoms are composed of even smaller components.

9–12 Structure and Properties of Atoms

- Atoms interact with one another by transferring or sharing electrons that are farthest from the nucleus. The outer electrons govern the chemical properties of the element.
- ★ Bonds between atoms are created when electrons are paired up by being transferred or shared. The atoms may be bonded together into molecules or crystalline solids.

★ Indicates a strong match between the ideas elicited by the probe and a national standard's learning goal.

Related Ideas in *Benchmarks for Science Literacy* (AAAS 1993 and 2008)

K–2 Structure of Matter

- Objects can be described in terms of the materials they are made of and their physical properties.

3–5 Structure of Matter

- Materials may be composed of parts that are too small to be seen without magnification.

6–8 Structure of Matter

- All matter is made up of atoms that are far too small to be seen directly through a microscope.
- ★ Atoms may link together in well-defined molecules or may be packed together in crystal patterns. Different arrangements of atoms into groups compose all substances and determine the characteristic properties of substances.
- ★ Atoms and molecules are perpetually in motion. In solids, the atoms are closely locked in position and can only vibrate.

9–12 Structure of Matter

- An enormous variety of biological, chemical, and physical phenomena can be explained by changes in the arrangement and motion of atoms and molecules.

Related Research

- Students of all ages show a wide range of beliefs about the nature and behavior of particles. For example, they attribute macroscopic properties to particles; do not accept the idea that there is empty space between particles, and have difficulty accepting the intrinsic motion of solids, liquids, and gases (AAAS 1993).
- Children frequently consider atoms of a solid to have all or most of the macro-properties they associate with the solid (Driver et al. 1994).
- Twenty-eight Australian 17-year-olds were interviewed in an Australian study conducted by Butts and Smith (1987) that focused on the formation of sodium chloride and use of the ball and stick model. The students referred to molecules of sodium chloride and stated that there were ionic bonds between the molecules.

Suggestions for Instruction and Assessment

- Use magnifiers to see the cuboidal shape of salt crystals. Challenge students to think what these cubes would look like at the microscopic level of the atom. Help them distinguish between the properties of the material (salt crystal) and the properties of the atoms. Just because a material has a certain shape does not mean the atoms have the same shape.
- A variety of materials, including ball and stick models, can be used to illustrate the ionic arrangement of the sodium and chloride atoms in table salt. However, make sure students do not think the sticks are actual physical structures between atoms.

★ Indicates a strong match between the ideas elicited by the probe and a national standard's learning goal.

- Clarify the difference between an ion and an atom.
- The PRISMS (Phenomena and Representations for Instruction of Science in Middle Schools) website at *http://prisms.mmsa.org* has a collection of reviewed web representations to help students visualize the atoms in a crystalline array. This website is part of the National Science Digital Library and can also be accessed through *http://nsdl.org*.

Related NSTA Science Store Publications, NSTA Journal Articles, NSTA SciGuides, NSTA SciPacks, and NSTA Science Objects

American Association for the Advancement of Science (AAAS). 2001. *Atlas of science literacy.* Vol. 1. (See "Atoms and Molecules" map, pp. 54–55.) Washington, DC: AAAS.

Logerwell, M., and D. Sterling. 2007. Fun with ionic compounds. *The Science Teacher* (Dec.): 27–33.

Robertson, W. 2007. *Chemistry basics: Stop faking it! Finally understanding science so you can teach it.* Arlington, VA: NSTA Press.

Related Curriculum Topic Study Guides

(Keeley 2005)
"Particulate Nature of Matter (Atoms and Molecules)"
"Solids"

References

American Association for the Advancement of Science (AAAS). 1993. *Benchmarks for science literacy.* New York: Oxford University Press.

American Association for the Advancement of Science (AAAS). 2008. Benchmarks for science literacy online. *www.project2061.org/publications/bsl/online*

Butts, B., and R. Smith. 1987. High school chemistry students' understanding of the structure and properties of molecular and ionic compounds. *Research in Science Education* 17 (1): 192–201.

Driver, R., A. Squires, P. Rushworth, and V. Wood-Robinson. 1994. *Making sense of secondary science: Research into children's ideas.* London: RoutledgeFalmer.

Keeley, P. 2005. *Science curriculum topic study: Bridging the gap between standards and practice.* Thousand Oaks, CA: Corwin Press.

National Research Council (NRC). 1996. *National science education standards.* Washington, DC: National Academy Press.

Ice Water

Christine put five ice cubes in a glass. After 20 minutes, most of the ice had melted to form "ice water." There were still some small pieces of ice floating in the water. Christine measured the temperature of the ice water then added five more ice cubes to the glass. She measured the temperature three minutes later. What do you predict happened to the temperature of the "ice water" three minutes after she added more ice?

A The temperature of the "ice water" increased.

B The temperature of the "ice water" decreased.

C The temperature of the "ice water" stayed the same.

Circle the answer that best matches your thinking. Explain what happens to the temperature of "ice water" when more ice is added.

_____

_____

_____

_____

_____

_____

_____

Ice Water

Teacher Notes

Purpose

The purpose of this assessment probe is to elicit students' ideas about temperature in the context of phases of matter. The probe is designed to find out if students recognize that the temperature of a substance does not change when two phases are present.

Related Concepts

energy, phase change, phases of matter, temperature, transfer of energy

Explanation

The best answer is C: The temperature of the ice water stayed the same. When an ice cube at −4°C is placed in a cup at room temperature (approximately 22°C), the surface of the solid cube absorbs thermal energy (sometimes called heat energy) from the surroundings. In

the solid phase, the molecules are being held in a relatively fixed position, and the energy is used to overcome the attractive forces between the water molecules. When sufficient energy is absorbed, the solid ice begins to melt. For water, this phase change occurs at 0°C.

During a phase change, the temperature of the system remains constant as long as two phases are present. In this example, the temperature will remain constant at 0°C while ice and water are both present. When more ice is added, the temperature will continue to remain constant at 0°C. Because the phenomenon is taking place in a cup surrounded by air at a temperature of 22°C, once all the ice melts, the energy in the system will result in an increased motion of the liquid molecules, and the water temperature will gradually rise until it reaches 22°C.

Temperature, heat, and thermal energy are related terms that are often confused. Temperature is the measure of the average kinetic energy of the particles that make up objects or materials. Heat is the amount of thermal energy that is transferred between two objects or materials due to a temperature difference. In other words, heat is thermal energy in transition as opposed to stored thermal energy. Thermal energy is the amount of internal kinetic and potential energy in an object or material.

Curricular and Instructional Considerations

Elementary Students

In the elementary grades, students use simple instruments to gather data. They learn to use thermometers to measure temperature. At this stage, their experience with temperature is observational. This is a time when students learn about phases of matter and changes from one phase to another. They can observe how temperature remains the same as a substance such as ice melts, thereby building an experiential foundation for explaining the relationship among heat, temperature, and phase change later on in middle school.

Middle School Students

In the middle grades, students begin to explain what happens during a phase change in terms of temperature patterns and begin to use ideas about energy transfer. They typically graph the change from ice to boiling water and analyze their graphs to understand that

the temperature remains the same as ice melts or water boils when two phases are present. Students in northern climates may draw on their everyday experience to connect the idea of the temperature of ice to weather-related phenomena, knowing that icy conditions happen when the temperature reaches 32°F (0°C). However, it may be counterintuitive to them that, after an extended period of time, the temperature of a sample of ice water (ice melting in water) sitting at room temperature will still be 0°C.

High School Students

Students at the high school level should be able to explain phase changes in terms of energy transfer. They should be able to extend their observations from ice to boiling water to predict what would happen if the steam continued to be heated. They should be encouraged to conceptually distinguish among heat, thermal energy, and temperature. However, memorizing the definition of these terms may not result in students being able to use them, such as in the example given in this probe.

Administering the Probe

This probe can be visually demonstrated to students by putting five ice cubes in a glass and letting the ice melt until there are small pieces of ice in the "ice water." Then add five more ice cubes and ask students to predict what will happen to the temperature of the ice water after a few minutes. With older students, you may wish to change the probe to include quantitative responses. Make sure students understand that

there are always some unmelted ice cubes present in the water. If the temperature were measured after all the ice had melted, the resulting temperature would be different. Note: Make sure students understand ice is not always 0°C. Ice can be colder than 0°C. 0°C is the melting point of ice.

Related Ideas in *National Science Education Standards* (NRC 1996)

K–4 Abilities Necessary to Do Scientific Inquiry

- Use simple equipment and tools to gather data and extend the senses.

K–4 Properties of Objects and Materials

- ★ Objects have many observable properties, including size, weight, shape, color, temperature, and the ability to react with other substances.
- Materials can exist in different states—solid, liquid, and gas. Some common materials, such as water, can be changed from one state to another by heating or cooling.

5–8 Transfer of Energy

- Energy is transferred in many ways.

9–12 Conservation of Energy and the Increase in Disorder

- Heat consists of random motion and the vibration of atoms, molecules, and ions.

The higher the temperature, the greater the atomic or molecular motion.

Related Ideas in *Benchmarks for Science Literacy* (AAAS 1993 and 2008)

Note: Benchmarks revised in 2008 are indicated by (R). New benchmarks added in 2008 are indicated by (N).

K–2 Constancy and Change

- ★ Things change in some ways and stay the same in some ways.

K–2 Scientific Inquiry

- Tools such as thermometers, magnifiers, rulers, or balances often give more information about things than can be obtained by just observing things without their help.

3–5 Constancy and Change

- ★ Some features of things may stay the same even when other features change.

3–5 Energy Transformation

- When warmer things are put with cooler ones, heat is transferred from the warmer ones to the cooler ones. (R)

6–8 Energy Transformation

- Thermal energy is transferred through a material by the collisions of atoms within the material. Over time, the thermal energy tends to spread out through a material and from one material to another if they are in contact. (R)

★ Indicates a strong match between the ideas elicited by the probe and a national standard's learning goal.

Related Research

- Driver and Russell (1982) carried out a study in which children were shown a beaker of ice with a thermometer reading of 0°C. The children were asked what would happen if more ice was added. Most of the 8-year-old children thought that the temperature would go up when more ice was added. Older children, up to age 14, thought that the temperature would decrease when more ice was added (Driver et al. 1994).

- Up to the age of 12, students are familiar with the term *temperature* and are able to use a thermometer to measure the temperature of objects or materials, but they have a fairly limited concept of the term. They rarely use temperature to describe the condition of an object (Erickson and Tiberghien 1985).

- Children do not recognize temperature as a physical parameter that can describe the condition of a material. They consider other criteria more pertinent for describing materials (Driver et al. 1994).

- Many 8- to 17-year-old students regard melting as a gradual process, unconnected with a particular temperature (Driver et al. 1994).

- Clough and Driver (1985) reported that students up to the age of 16 think of cold as "an entity which, like heat, has the properties of a material substance." They do not necessarily think of hot and cold as part of the same continuum, but instead they think of cold as the opposite of heat (Driver et al. 1994).

- Students may use the intuitive rule "more A, more B" to reason what happens when you have more ice cubes (Stavy and Tirosh 2000). Using this rule, students may think that when there is more ice, there will be "more cold."

Suggestions for Instruction and Assessment

- This probe can be used as a P-E-O-E probe—Predict, Explain, Observe, Explain (again). Have students predict what would happen and explain the reasons for their predictions. Then have students test their ideas and observe what happens to the temperature. If their observations do not match their predictions, students are encouraged to come up with revised explanations.

- Use ABC—"activity before concept" (Eisenkraft 2006)—to provide an opportunity for students to experience the phenomenon and collect and analyze temperature data before presenting them with the concept of temperature during a phase change. Having students observe and construct their own explanations first before being presented with the formal scientific concepts is a more powerful way for students to learn.

- Use a real-life phenomenon to demonstrate that temperature remains constant during a phase change. Obtain an extremely large sample of crushed ice or, in northern climates during winter, fill a large bucket with snow from the school yard. Record the temperature of the snowy slush in the bucket throughout the day.

- Have students use phase-change graphs to analyze patterns and notice that when two phases are present (e.g., water in the liquid and solid form), the temperature remains the same.

- Be sure to explicitly develop the generalization that constant temperature during a phase change (melting, freezing, boiling, condensation) is true for all pure substances, not just water.

- Do not introduce the difference between heat and temperature in the context of this probe until students are ready to understand this difference. Up through late middle school, it may suffice to keep this idea at an observational level and hold off on explanations until students are ready.

- With older students, link the idea of energy transfer to phase change. Ask students to discuss why the temperature of the water remains constant and then increases as soon as all of the ice melts. What was happening to the energy while there was still ice in the system?

Related NSTA Science Store Publications, NSTA Journal Articles, NSTA SciGuides, NSTA SciPacks, and NSTA Science Objects

Konicek-Moran, R. 2008. *Everyday science mysteries: Stories for inquiry-based science teaching.* Arlington, VA: NSTA Press.

Robertson, W. 2002. *Energy: Stop faking it! Finally understanding science so you can teach it.* Ar-

lington, VA: NSTA Press.

National Science Teachers Association. 2006. NSTA Energy SciPak. Online at *www.nsta. org/store/product_detail.aspx?id-id=10.2505/7/ SCB-GO.3.1*

Related Curriculum Topic Study Guides

(Keeley 2005)

"Heat and Temperature"

"States of Matter"

References

American Association for the Advancement of Science (AAAS). 1993. *Benchmarks for science literacy.* New York: Oxford University Press.

American Association for the Advancement of Science (AAAS). 2008. Benchmarks for science literacy online. *www.project2061.org/publications/ bsl/online*

Driver, R., and T. Russell. 1982. *An investigation of the ideas of heat, temperature, and change in state of children aged between 8 and 14 years.* Leeds, England: Centre for Studies in Science and Mathematics Education, University of Leeds.

Driver, R., A. Squires, P. Rushworth, and V. Wood-Robinson. 1994. *Making sense of secondary science: Research into children's ideas.* London: RoutledgeFalmer.

Eisenkraft, A. 2006. ABC—Activity before concept. Presented at the National Science Teachers Association National Conference, Saint Louis.

Engel Clough, E., and R. Driver. 1985. Secondary students' conception of the conduction of heat: Bringing together scientific and personal views.

Physics Education 20: 176–82.

Erickson, G., and A. Tiberghien. 1985. Heat and temperature. In *Children's ideas in science*, eds. R. Driver, E. Guesne, and A. Tiberghien, 52–84. Milton Keynes, England: Open University Press.

Keeley, P. 2005. *Science curriculum topic study: Bridging the gap between standards and practice.*

Thousand Oaks, CA: Corwin Press.

National Research Council (NRC). 1996. *National science education standards.* Washington, DC: National Academy Press.

Stavy, R., and D. Tirosh. 2000. *How students (mis-) understand science and mathematics: Intuitive rules.* New York: Teachers College Press.

Warming Water

Two friends put a bowl of very cold water outside on a hot sunny day. The Sun warmed the water. They wondered about the energy of the water. This is what they thought:

Ted: "The very cold water had energy. The Sun provided additional energy to warm the water."

Ambra: "The very cold water did not have energy. The energy in the warm water came from the Sun."

Which friend has the best idea? Explain why you agree with one friend and not the other.

_____

_____

_____

_____

_____

_____

_____

_____

_____

_____

Warming Water

Teacher Notes

Purpose

The purpose of this assessment probe is to elicit students' ideas about thermal energy. The probe is designed to find out whether students think cold things can have energy.

Related Concepts

energy, heat, temperature, thermal energy, transfer of energy

Explanation

The best answer is Ted's: "The very cold water had energy. The Sun provided additional energy to warm the water." Under ordinary conditions, all objects, materials, and substances "possess" an internal energy called thermal energy. Even very cold objects like ice cubes have thermal energy. The thermal energy of the water is the total of all the kinetic energy (due to molecular motion) and potential energy (because of relative position or shape of the molecules) in the bowl of water. Molecules are in constant motion, even in very cold water. The difference is that the molecules in the cold water move slower than the molecules in warm water. The cold water has less thermal energy before it is warmed by the Sun, but nevertheless it still has some thermal energy even at a cold temperature. The Sun warms the water by transferring energy from the Sun to the cold water. The gain in energy changes the amount of thermal energy the water has. As the water molecules gain energy, they move faster and the water temperature increases. Both the cold and warm water have energy; however, the bowl of water has more thermal energy when it is warm than when it is cold.

Temperature, heat, and thermal energy (sometimes referred to with younger students as heat energy) are related terms that are often confused. Temperature is the measure of the average kinetic energy of the particles that make up objects or materials. Heat is the amount of thermal energy that is transferred between two objects or materials because of a temperature difference. In other words, heat is thermal energy in transition as opposed to "stored" thermal energy. Thermal energy is the amount of internal kinetic and potential energy in an object or material. For example, the thermal energy of a massive iceberg will be much larger than that of a cup of boiling water, despite its much lower temperature, simply because it has more molecules.

Curricular and Instructional Considerations

. . . .

Elementary Students

In the elementary grades, energy is a complex concept. Even though students have heard the word *energy*, investing a lot of time and effort in developing formal energy concepts should wait until middle school (AAAS 1993). Young children have intuitive notions about energy (e.g., energy gets things done) that teachers can build on without getting into details of formal energy concepts. At this level, students can talk about energy but should not be expected to define it. One aspect of energy that can be developed at this age is the idea of heat. Students can observe how heat spreads from one object or place to another and can consider ways to increase or

decrease the spreading of heat. They should also be encouraged to wonder where the energy comes from that makes things happen.

Middle School Students

In the middle grades, students are introduced to energy through energy transformations and transfer. At this level, they describe various forms of energy including chemical, thermal, electrical, mechanical, electromagnetic, gravitational potential, elastic potential, and kinetic energy. They trace where forms of energy come from and where the energy goes. By grade 8, they should transition from using the commonly used term *heat energy* to describe an object's internal energy to using the scientific term *thermal energy*. However, students at this level still have difficulty distinguishing among heat, thermal energy, and temperature.

High School Students

Students at the high school level take the variety of energy forms described in middle school and begin to see that they fall into a few basic types: kinetic energy, potential energy, or energy contained by a field such as electromagnetic waves. An increased understanding of temperature, atoms, and molecules helps them relate the motion of molecules, as well as the position and shape of molecules, to the concept of thermal energy. They should have a variety of opportunities to investigate how energy interacts with matter either by losing or by gaining energy. At this level, students are now able to see how powerful energy ideas are in explaining phenomena. Even though they

are introduced to the theoretical idea of "absolute zero" as the temperature at which molecular motion ceases, some high school students may believe that cold water has energy but not accept the idea that ice has energy.

Administering the Probe

As an assessment used to inform instruction, this probe is best used with middle and high school students to find out their ideas about what they commonly refer to as *heat energy* before formally encountering the term *thermal energy*. You might consider including the temperature of the water. For example, explain that the cold water is 4°C and warms up to 30°C in the sun. A variation of this probe that would include phase change is to use an example of a glass of ice cubes at 0°C and the same glass after the ice cubes have melted in the sun. Although some students may think the water has energy, they may think that a frozen solid does not have energy. Some students may also think a temperature of 0°C may mean zero energy.

Related Ideas in *National Science Education Standards* (NRC 1996)

. . . .

K–4 Properties of Objects and Materials

• Objects have many observable properties, including temperature. Temperature can be measured using a thermometer.

5–8 Transfer of Energy

★ Energy is a property of many substances and is associated with heat, light, electricity, mechanical motion, sound nuclei, and the nature of a chemical. Energy is transferred in many ways.

• The Sun is a major source of energy for changes on the Earth's surface. The Sun loses energy by emitting light. A tiny fraction of that light reaches Earth, transferring energy from the Sun to the Earth.

9–12 Conservation of Energy and the Increase in Disorder

★ Heat consists of random motion and the vibrations of atoms, molecules, and ions. The higher the temperature, the greater the atomic or molecular motion.

9–12 Interactions of Energy and Matter

• Waves (including light waves) have energy and can transfer energy when they interact with matter.

Related Ideas in *Benchmarks for Science Literacy* (AAAS 1993 and 2008)

. . . .

Note: Benchmarks revised in 2008 are indicated by (R). New benchmarks added in 2008 are indicated by (N).

K–2 Energy Transformation

• The Sun warms the land, air, and water.

★ Indicates a strong match between the ideas elicited by the probe and a national standard's learning goal.

3–5 Energy Transformation

- Things that give off light often also give off heat.
- When warmer things are put with cooler ones, heat is transferred from the warmer ones to the cooler ones. (R)

6–8 Energy Transformation

- Thermal energy is associated with the temperature of an object. (R)
- ★ Energy can be transferred from one system to another (or from a system to its environment) in different ways: thermally, when a warmer object is in contact with a cooler one … and by electromagnetic waves. (R)
- Thermal energy is transferred through a material by the collisions of atoms within the material. Over time, the thermal energy tends to spread out through a material and from one material to another if they are in contact. Thermal energy can also be transferred by means of currents in air, water, or other fluids. In addition, some thermal energy in all materials is transformed into light energy and radiated into the environment by electromagnetic waves; that light energy can be transformed back into thermal energy when the electromagnetic waves strike another material. As a result, a material tends to cool down unless some other form of energy is converted to thermal energy in the material. (R)

9–12 Energy Transformation

- ★ Thermal energy is a system that is associated with the disordered motions of its atoms or molecules.
- Many forms of energy can be considered to be either *kinetic energy*, which is the energy of motion, or *potential energy*, which depends on the separation between mutually attracting or repelling objects. (N)

Related Research

- Students' meanings for energy, both before and after traditional instruction, are considerably different from the scientific meaning of energy (AAAS 1993).
- Several studies have found students to have an anthropomorphic view of energy. Students tend to associate energy with living things—in particular, human beings—or with objects that were treated as if they had human characteristics. They suggest that energy is needed to live or be active, or they relate it to fitness and strength (Driver et al. 1994).
- Many students think of energy as a substance that is stored in some objects but not in others (Driver et al. 1994).
- Distinguishing between concepts of heat and temperature is difficult for most children. They tend to view temperature as the mixture of hot and cold inside an object or simply the measure of the amount of thermal energy (often referred to as "heat energy" with younger children) possessed by an object with no distinction between the temperature of an object and its ther-

---

★ Indicates a strong match between the ideas elicited by the probe and a national standard's learning goal.

mal energy. Studies show that students ages 10–16 tend to think there is no difference between heat and temperature (Driver et al. 1994).

- Watts and Gilbert (1985) found that it was common for 14- to 17-year-olds to associate heat only with warm and hot bodies.

Suggestions for Instruction and Assessment

- Although the word *energy* is familiar to students, energy is far more difficult to teach than is often recognized. In science, energy is an abstract, mathematical idea. It is hard to define energy or even to explain clearly what is meant by the word. This means that in order to communicate the scientific idea of energy to students, teachers must first simplify it—but still ensure that what is taught is clear and useful and provides a sound basis for developing a fuller understanding later (Millar 2005).

- The word *energy* is widely used in everyday contexts, including many that appear "scientific." However, *energy* is used in a way that is less precise than its scientific meaning and that differs from its scientific meaning in certain respects. This means that teachers have to be very careful to disentangle the everyday use of the word *energy* from its scientific use in order to both keep teachers' and students' own ideas clear and avoid teaching a potentially confusing mixture of the two (Millar 2005).

- Ask students to write three to four sentences showing how they would use the word *energy*. Their responses often reveal what types of things they associate energy with.

- Be aware that teaching students to distinguish between heat and temperature is not a "quick fix." Researchers have found that long-term teaching interventions are required for upper middle school students to start differentiating between the two concepts (Linn and Songer 1991).

- A suggested progression of ideas on understanding energy begins in elementary grades with observable patterns of phenomena involving heat transfer. In middle school, teaching and learning move to descriptions of various forms of energy and examples of energy transfer and transformation. High school learning culminates with an understanding of energy conservation and dissipation as thermal energy (AAAS 2007).

- When teaching ideas that are related to energy transfer, help students define the system under consideration. Because systems often interact with their environment, students should practice keeping track of what enters or leaves the system (AAAS 2007).

Related NSTA Science Store Publications, NSTA Journal Articles, NSTA SciGuides, NSTA SciPacks, and NSTA Science Objects

American Association for the Advancement of Science (AAAS). 2007. *Atlas of science literacy.* Vol. 2. (See "Energy Transformations," pp. 24–25.) Washington, DC: AAAS.

National Science Teachers Association. 2006. NSTA Energy SciPack. Online at *http://learningcenter. nsta.org/product_detail.aspx?id=10.2505/6/ SCP-OCW.0.1*

Robertson, W. 2002. *Energy: Stop faking it! Finally understanding science so you can teach it.* Arlington, VA: NSTA Press.

Robertson, W. 2007. Science 101: What exactly is energy? *Science & Children* (Mar.): 62–63.

Related Curriculum Topic Study Guides

(Keeley 2005)
"Energy"
"Heat and Temperature"

References

American Association for the Advancement of Science (AAAS). 1993. *Benchmarks for science literacy.* New York: Oxford University Press.

American Association for the Advancement of Science (AAAS). 2007. *Atlas of Science Literacy.* Vol. 2. (See "Energy Transformations," pp. 24–25.) Washington, DC: AAAS.

American Association for the Advancement of Science (AAAS). 2008. Benchmarks for science literacy online. *www.project2061.org/publications/ bsl/online*

Driver, R., A. Squires, P. Rushworth, and V. Wood-Robinson. 1994. *Making sense of secondary science: Research into children's ideas.* London: RoutledgeFalmer.

Keeley, P. 2005. *Science curriculum topic study: Bridging the gap between standards and practice.* Thousand Oaks, CA: Corwin Press.

Linn, M., and N. Songer. 1991. Teaching thermodynamics to middle school students: What are appropriate cognitive demands? *Journal of Research in Science Teaching* 28 (10): 885–918.

Millar, R. 2005. *Teaching about energy.* Department of Educational Studies Research Paper 2005/11. York, UK: The University of York.

National Research Council (NRC). 1996. *National science education standards.* Washington DC: National Academy Press.

Watts, D., and J. Gilbert. 1985. *Appraising the understanding of science concepts: Heat.* Surrey, UK: Department of Educational Studies, University of Surrey, Guildford.

Standing on One Foot

Maria stood on her bathroom scale with two feet. She read her weight on the scale. She then lifted one foot. Circle what you think happened to the reading on the scale when she stood on one foot.

A It showed an increase in weight.

B It showed a decrease in weight.

C Her weight stayed the same.

Explain your thinking. What "rule" or reasoning did you use to select your answer?

_____

_____

_____

_____

_____

_____

_____

_____

_____

_____

_____

_____

_____

Standing on One Foot

Teacher Notes

Purpose

The purpose of this assessment probe is to elicit students' ideas about weight and pressure. The probe is designed to determine whether students think their weight changes when the force exerted per unit area (pressure) on a scale changes.

Related Concepts

force, gravity, pressure, weight

Explanation

The best answer is C: Her weight stayed the same. Weight is the force of gravity acting on an object. Regardless of whether you stand on two feet or one foot, the force of gravity acting on your body as you stand on a bathroom scale is the same. When you stand on two feet, the force is distributed over a wider area (the total area covered by the two soles of your feet). When you lift one foot, the same force is distributed over a smaller area (the area covered by the sole of one of your feet). Pressure changes as the constant weight of the body is distributed over different areas. Because pressure is described as force per unit area ($P = F \div A$), as the area covered by the body on the scale decreases by lifting one foot, the pressure increases. Although the pressure increases, the weight remains constant.

Curricular and Instructional Considerations

Elementary Students

In the elementary grades, students use simple instruments to gather data. They learn to measure weight and mass using various types of scales and pan balances. At this stage, weight

is an observational property that they use to describe objects and materials. Elementary students are not expected to know that weight is caused by the force of gravity. However, they should be able to observe that the weight of an object stays the same in the same location on Earth.

Middle School Students

At the middle school level, students should begin to distinguish between weight and mass. They develop an understanding that the force of the Earth's gravity affects the weight of an object. They often confuse pressure with weight. This is a time when their use of mathematics (area and proportionality) can be used to explain why their weight doesn't change when different amounts of their body are in contact with the scale.

High School Students

In high school, students develop more sophisticated notions of weight, mass, and pressure. However, they may still revert to naive ideas about weight.

Administering the Probe

You can demonstrate the context for this probe by bringing in a bathroom scale, standing on it with two feet, then lifting one foot. Make sure students do not see the reading on the scale as you demonstrate.

Related Ideas in *National Science Education Standards* (NRC 1996)

. . . .

K–4 Abilities Necessary to Do Scientific Inquiry

★ Use simple equipment and tools to gather data and extend the senses.

K–4 Properties of Objects and Materials

★ Objects have many observable properties, including size, weight, shape, color, and the ability to react with other substances. Those properties can be measured using tools, such as rulers, balances, and thermometers.

5–8 Abilities Necessary to Do Scientific Inquiry

★ Use appropriate tools and techniques to gather, analyze, and interpret data.

9–12 Motions and Forces

• Gravitation is a universal force that each mass exerts on any other mass.

Related Ideas in *Benchmarks for Science Literacy* (AAAS 1993 and 2008)

. . . .

Note: Benchmarks revised in 2008 are indicated by (R). New benchmarks added in 2008 are indicated by (N).

K–2 Scientific Inquiry

• People can often learn about things around them by just observing those things care-

---

★ Indicates a strong match between the ideas elicited by the probe and a national standard's learning goal.

fully, but sometimes they can learn more by doing something to the things and noting what happens.

- Tools such as thermometers, magnifiers, rulers, and balances often give more information about things than can be obtained by just observing things unaided.

K–2 Structure of Matter

- Objects can be described in terms of their properties.

3–5 Forces of Nature

- The Earth's gravity pulls any object on or near the Earth toward it without touching it. (R)

6–8 Forces of Nature

- Every object exerts gravitational force on every other object.

9–12 Forces of Nature

- Gravitational force is an attraction between masses. The strength of the force is proportional to the masses and weakens rapidly with increasing distance between them.

Related Research

- The idea that the weight of an object is a force—the force of gravity on that object—does not appear to be a firmly held idea among secondary students (Driver et al. 1994).
- There is not a lot of research on students' ability to distinguish between weight and pressure in relation to the surface area

of an object in contact with a weighing instrument, such as a bathroom scale. Our field tests with elementary and middle school students showed that many students believed that the weight increases when there is less surface area touching the scale because "it presses down harder." Some students also believed the weight would decrease because the amount of force pushing down is less when one foot is lifted. These ideas were less prevalent with high school students, although some students still believed standing on one foot would press down harder and increase the weight.

Suggestions for Instruction and Assessment

- This probe lends itself to an inquiry investigation. Have students commit to a prediction, explain their reasoning that supports their prediction and then test it. When students find their observation does not match their prediction, encourage them to discuss their ideas and seek information to support a new explanation.
- With younger children, have them weigh a variety of objects by placing them on their side, top, bottom, or in other configurations, observing how the weight stays the same. Be sure to place the object in the center of the weighing pan or device each time.
- Challenge students with other examples illustrating the difference between weight and pressure. For example, imagine that you had two different pairs of shoes that

weighed the same. One was a pair of high spiked heels and the other was a pair with flat soles. When you put the high spiked heels on and stood on soft ground, the heel sunk down into the ground. When you put the flat sole on, you stayed on top of the ground. What changed, your weight or the shoes? Why did this change in the shoes affect how you stood on the soft ground?

- After developing an operational definition of pressure, have students compare the pressure exerted by one foot on the scale with two feet on the scale. Have students trace their feet on square grid paper and count up the number of square units covered by one versus two feet. Calculate the pressure by dividing their weight by the square units.

Related NSTA Science Store Publications, NSTA Journal Articles, NSTA SciGuides, NSTA SciPacks, and NSTA Science Objects

American Association for the Advancement of Science (AAAS). 2001. *Atlas of science literacy.* Vol. 1. (See "Gravity," pp. 42–43.) Washington, DC: AAAS.

Nelson, G. 2004. What is gravity? *Science & Children* (Sept.): 22–23.

Robertson, W. 2002. *Force and motion: Stop faking it! Finally understanding science so you can teach it.* Arlington, VA: NSTA Press.

Related Curriculum Topic Study Guides

(Keeley 2005)

"Observation, Measurement, and Tools"
"Gravitational Force"

References

American Association for the Advancement of Science (AAAS). 1993. *Benchmarks for science literacy.* New York: Oxford University Press.

American Association for the Advancement of Science (AAAS). 2008. Benchmarks for science literacy online. *http://www.project2061.org/publications/bsl/online*

Driver, R., A. Squires, P. Rushworth, and V. Wood-Robinson. 1994. *Making sense of secondary science: Research into children's ideas.* London: RoutledgeFalmer.

Keeley, P. 2005. *Science curriculum topic study: Bridging the gap between standards and practice.* Thousand Oaks, CA: Corwin Press.

National Research Council (NRC). 1996. *National science education standards.* Washington, DC: National Academy Press.

Magnets in Water

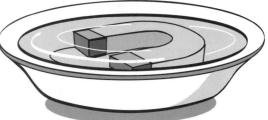

Four friends were wondering if a magnet could pick up steel paper clips in water. This is what they said:

Nate: "I think magnets and paper clips need to be in air. If both the magnets and paper clips are in water, they won't attract."

Amy: "I think magnets need to be in the air, but it doesn't matter if the paper clip is. Magnets can attract paper clips covered with water."

Steve: "I don't think air makes a difference. I think magnets will attract paper clips when both are underwater."

Leah: "I don't think air makes a difference. However, when magnets are in water, they work the opposite way. The paper clips will be repelled by the magnet."

Which friend do you agree with and why? Explain your thinking about how magnets work.

_____

_____

_____

_____

_____

_____

_____

_____

Magnets in Water

Teacher Notes

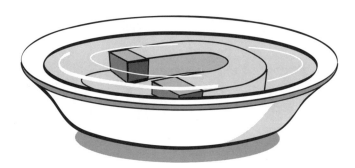

Purpose

The purpose of this assessment probe is to elicit students' ideas about magnetism. The probe is specifically designed to determine whether students believe air is necessary for magnets to work.

Related Concept

magnetism

Explanation

The best answer is Steve's: "I don't think air makes a difference. I think magnets will attract paper clips when both are underwater." Magnetism is a force that can work through a gas, a liquid, and even a solid (e.g., nonmagnetic materials such as paper, wood, aluminum foil, tape, and plastic). Just as electricity moves through some materials better than others, magnetism moves with ease through some materials and has more difficulty passing through other materials. Although most peoples' experiences with magnets happen in an environment in which the magnet is surrounded by air, magnets also work underwater and in other gaseous environments, such as in carbon dioxide and helium. Magnets also work in environments without an atmosphere or air. For example, a magnet on the Moon or a magnet in a bell jar with all the air removed will attract iron objects.

Curricular and Instructional Considerations

Elementary Students

In the elementary grades, magnets provide students with multiple opportunities to engage in

inquiry while developing the idea that forces can act without directly touching an object. Students' experiences with magnets are mostly observational. Explaining the specifics of how magnets work should wait until students develop a deeper understanding of forces in middle school.

Middle School Students

In the middle grades, students combine their knowledge of magnets with their understanding of electric current. At this level, students have developed ideas about gravity and electric charge that may interfere with ideas about magnetism.

High School Students

At the high school level, students develop more sophisticated ideas about electromagnetism. However, early misconceptions related to magnets and their effect in air may still persist.

Administering the Probe

This probe is appropriate at all grade levels. You might consider using a prop by first showing how magnets pick up paper clips when both are in air and then putting a magnet in water and asking students what they think would happen if the paper clips were placed near the magnet. This probe can lead to a lively discussion of students' ideas.

Related Ideas in *National Science Education Standards* (NRC 1996)

. . . .

K–4 Properties of Objects and Materials

- Objects have many observable properties, including size, weight, shape, color, temperature, and the ability to react with other substances.

K–4 Light, Heat, Electricity, and Magnetism

★ Magnets attract and repel each other and certain kinds of materials.

9–12 Motions and Forces

- Electricity and magnetism are two aspects of a single electromagnetic force.

Related Ideas in *Benchmarks for Science Literacy* (AAAS 1993 and 2008)

Note: Benchmarks revised in 2008 are indicated by (R). New benchmarks added in 2008 are indicated by (N).

K–2 Forces of Nature

- Magnets can be used to make some things move without being touched.

3–5 Forces of Nature

★ Without touching them, a magnet pulls on all things made of iron and either pushes or pulls on other magnets.

6–8 Forces of Nature

- Electric currents and magnets can exert a force on each other.

★ Indicates a strong match between the ideas elicited by the probe and a national standard's learning goal.

9–12 Forces of Nature

*   Magnetic forces are very closely related to electric forces and are thought of as different aspects of a single electromagnetic force. Moving electrically charged objects produces magnetic forces and moving magnets produces electric forces. (R)

Related Research

*   Research has shown that some students are inclined to link gravity with magnetism (Driver et al. 1994). If they believe gravity is necessary for magnets to work and also believe that gravity has no or less of an effect under water, they may believe magnets will not attract objects in water.
*   Barrow (1987) investigated students' awareness of magnets and magnetism across age ranges and found that they were aware of magnets through their everyday experiences of sticking objects to refrigerators with magnets. However, before instruction, few students could offer explanations of magnetism, especially in terms of forces and how magnets work (Driver et al. 1994).
*   Bar and Zinn (1989) sampled 98 students ages 9–14 and found that 40% believed that a medium (air) was necessary in order for magnets to have an effect on objects. Twenty percent of these students also made a link between gravity and magnetism (Driver et al. 1994).

Suggestions for Instruction and Assessment

*   This probe can be used with the P-E-O (Predict, Explain, Observe) strategy (Keeley 2008). Have students make a prediction, explain the reasons for their prediction, and test their ideas. If their observations do not match their predictions, challenge them to construct a new explanation to fit their observations.
*   Consider probing students' ideas about magnets related to their ideas about gravity. Do students believe gravity is necessary in order for magnets to work? Do they link the idea that air is necessary for gravity and thus link magnetism to gravity as well?
*   Barrow's (1987) study considered that teaching about magnets might dissociate students from their everyday awareness of magnetism. Barrow suggested that teaching approaches that draw on everyday experience and focus on uses of magnets may be effective in helping students understand magnetism.
*   Providing opportunities for students to test and observe how magnetism passes through solid materials such as paper, plastic, and aluminum foil may help support the notion that magnetism also passes through a liquid such as water.
*   Challenge students to think about whether magnets work in the absence of air and come up with a way to test it. One way is to suspend a magnet attracted to a paper clip in a bell jar. If students think air is needed in order for the paper clip to remain

attracted to the magnet, have them observe whether the paper clip will drop.

Related NSTA Science Store Publications, NSTA Journal Articles, NSTA SciGuides, NSTA SciPacks, and NSTA Science Objects

American Association for the Advancement of Science (AAAS). 2007. *Atlas of science literacy.* Vol. 2. (See "Electricity and Magnetism" map, pp. 26–27.) Washington, DC: AAAS.

Ansberry, K., and E. Morgan. 2007. *More picture-perfect science lessons: Using children's books to guide inquiry, K–4.* (See "That Magnetic Dog," pp. 123–129.) Arlington, VA: NSTA Press.

Kur, J., and M. Heitzman. 2008. Attracting student wonderings. *Science & Children* (Jan.): 28–32.

Robertson, W. 2005. *Electricity and magnetism: Stop faking it! Finally understanding science so you can teach it.* Arlington, VA: NSTA Press.

> **Related Curriculum Topic Study Guide**
> (Keeley 2005)
> "Magnetism"

References

American Association for the Advancement of Science (AAAS). 1993. *Benchmarks for science literacy.* New York: Oxford University Press.

American Association for the Advancement of Science (AAAS). 2008. Benchmarks for science literacy online. *www.project2061.org/publications/bsl/online*

Bar, V. and B. Zinn. 1989. *Does a magnet act on the moon?* Scientific Report, The Amos de Shalit Teaching Centre. Jerusalem, Israel: Hebrew University.

Barrow, L. 1987. Magnet concepts and elementary students' misconceptions. In *Proceedings of the second international seminar on misconceptions and educational strategies in science and mathematics,* ed. J. Novak, 3: 17–22. Ithaca, NY: Cornell University.

Driver, R., A. Squires, P. Rushworth, and V. Wood-Robinson. 1994. *Making sense of secondary science: Research into children's ideas.* London: RoutledgeFalmer.

Keeley, P. 2005. *Science curriculum topic study: Bridging the gap between standards and practice.* Thousand Oaks, CA: Corwin Press.

Keeley, P. 2008. *Science formative assessment: 75 practical strategies for linking assessment, instruction, and learning.* Thousand Oaks, CA: Corwin Press.

National Research Council (NRC). 1996. *National science education standards.* Washington, DC: National Academy Press.

Is It a Model?

Below are listed things that students might do in a science class. Check off the things that are examples of using a model.

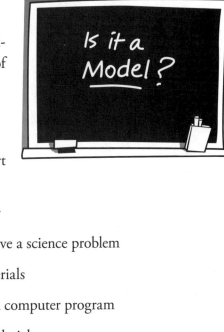

_____ **A** building a paper airplane

_____ **B** making an analogy (for example, the heart is like a pump)

_____ **C** observing a bird's behavior at a bird feeder

_____ **D** developing a mathematical equation to solve a science problem

_____ **E** making a plant cell out of household materials

_____ **F** analyzing whale migration patterns with a computer program

_____ **G** building and testing a bridge made of toothpicks

_____ **H** drawing an electrical circuit

_____ **I** forming a mental image of molecules in the liquid state

_____ **J** demonstrating the day/night cycle with a globe and flashlight

_____ **K** dissecting a cow's bone

_____ **L** watching a computer simulation of a hurricane

_____ **M** going on a field trip to the Grand Canyon

_____ **N** graphing the speed of a car

_____ **O** watching a live video of an active volcano

_____ **P** making a replica of a human heart out of clay

_____ **Q** looking at blood cells under a microscope

Explain your thinking. How did you decide whether something is a model?

_____

Is It a Model?

Teacher Notes

Purpose

The purpose of this assessment probe is to elicit students' ideas about models. The probe is designed to find out whether students recognize that models can take a variety of forms besides physical replicas.

Related Concepts

model

Explanation

The best answer is that all but C, K, M, O, and Q are examples of using models. Models are representations of objects, processes, or phenomena that look like, function like, describe, or explain the real thing. They are often simplified versions of the real object and help us to understand how things work. C, K, M, O, and Q all involve the real

object or process being studied and thus are not models.

There are many types of models. Physical models can be made from similar or different materials and be smaller or larger in size with the same proportional scale (paper airplane, cell made from household materials, toothpick bridge, globe and flashlight to show day/night cycle, clay heart). Drawings and illustrations are models that help us understand real objects, places, or processes (electrical circuit drawing). Conceptual models help us make sense of an unfamiliar, complex, or abstract idea (mental image, analogy). Drawings are also used to represent conceptual models, such as drawing how one thinks molecules are arranged in water. Mathematical models use a relationship that represents the behavior or properties of an object or system (e.g., mathematical equa-

tion to solve a problem, graph). Computers can be used for simulating or mathematically representing a system that may be difficult to observe and analyze in real time (e.g., analyzing migration patterns, hurricane simulation). Models play a crucial role in science. Scientists use models as thinking tools to develop explanations for phenomena and to make predictions about phenomena. Because models are so important to scientists, they evaluate them on an ongoing basis. They consider how well models explain or predict past observations, and they look at new observations that have been collected. They evaluate models to make sure they are consistent with what is understood about nature. They look to determine how well models make predictions, and they test their accuracy. Scientists also are well aware of the limitations of their models. They may be appropriate only under certain conditions. Scientists also frequently revise their models if they fail to explain or predict well.

Curricular and Instructional Considerations

Elementary Students

In the elementary grades, students talk about how the things they play with relate to real-world objects. As they progress, students begin to talk about limitations and make changes to their physical models (e.g., modify wheels, use different materials). They discuss how mathematical concepts can be used to represent natural phenomena and use analogies to make sense of complex ideas. Students com-

pare their models with the real thing, formulate their own models to explain things they can not observe directly, and test their models as more information is obtained, thus building an understanding of how science works.

Middle School Students

In the middle grades, students have a greater general knowledge of mathematics, objects, and processes. Students become familiar with phenomena and systems in the world around them through a variety of direct experiences, and models are used explicitly to build scientific understanding. Computers are used for graphing and simulations that calculate and depict what happens when variables are changed. Students use conceptual models to pose hypothetical questions, and changes in spatial or temporal scale of physical models become increasingly sophisticated. Models that reveal patterns or trends are used to develop generalizations, and the process of evaluating models strengthens justification skills.

High School Students

Students at the high school level learn how to create and use models in a variety of contexts, and much emphasis is placed on mathematical modeling. Students continue to develop generalizations through discussion of models and use the graphic capabilities of computers to design and test models that simulate complicated processes. Students encounter the ideas that there is no one "true" model and that scientists may not have the best model because not enough information is available. Students

test models by comparing predictions with actual observations.

Administering the Probe

This probe is best used as is at the middle and high school level, particularly if students have been previously exposed to the word *model* or its use. Remove any answer choices students might not be familiar with. The probe can also be modified as a simpler version for students in grades 3–5 by reducing the number of choices, specifically leaving out some of the more complex and unfamiliar choices.

Related Ideas in *National Science Education Standards* (NRC 1996)

· ·

K–12 Unifying Concepts and Processes—Evidence, Models, and Explanation

- Models are tentative schemes or structures that correspond to real objects, events, or classes of events and that have explanatory power.
- Models help scientists and engineers understand how things work.
- ★ Models take many forms, including physical objects, plans, mental constructs, mathematical equations, and computer simulations.

Related Ideas in *Benchmarks for Science Literacy* (AAAS 1993 and 2008)

· ·

Note: Benchmarks revised in 2008 are indicated by

(R). New benchmarks added in 2008 are indicated by (N).

K–2 Models

- Many of the toys children play with are like real things only in some ways. They are not the same size, are missing many details, or are not able to do all of the same things.
- A model of something is different from the real thing but can be used to learn something about the real thing.
- One way to describe something is to say how it is like something else.

3–5 Models

- A model of something is similar to, but not exactly like, the thing being modeled. Some models are physically similar to what they are representing, but others are not. (N)
- ★ Geometric figures, number sequences, graphs, diagrams, sketches, number lines, maps, and stories can be used to represent objects, events, and processes in the real world, although such representations can never be exact in every detail. (R)
- Models are very useful for communicating ideas about objects, events, and processes. When using a model to communicate about something, it is important to keep in mind how it is different from the thing being modeled. (N)

6–8 Models

- Models are often used to think about processes that happen too slowly, too quickly, or on too small a scale to observe directly,

---

★ Indicates a strong match between the ideas elicited by the probe and a national standard's learning goal.

or that are too vast to be changed deliberately, or that are potentially dangerous.

★ Mathematical models can be displayed on a computer and then modified to see what happens.

• Different models can be used to represent the same thing. What kind of a model to use and how complex it should be depends on its purpose. The usefulness of a model may be limited if it is too simple or if it is needlessly complicated. Choosing a useful model is one of the instances in which intuition and creativity come into play in science, mathematics, and engineering.

9–12 Models

★ A mathematical model uses rules and relationships to describe and predict objects and events in the real world. (R)

• Computers have greatly improved the power and use of mathematical models by performing computations that are very long, very complicated, or repetitive. Therefore computers can show the consequences of applying complex rules or of changing the rules. The graphic capabilities of computers make them useful in the design and testing of devices and structures and in the simulation of complicated processes.

• The behavior of a physical model cannot ever be expected to represent the full-scale phenomenon with complete accuracy, not even in the limited set of characteristics being studied. The inappropriateness of a model may be related to differences

between the model and what is being modeled. (N)

Related Research

• Students may think that models are physical copies of the real thing, failing to recognize models as conceptual representations (AAAS 1993).

• In the lower elementary grades, students have some understanding of how models are used, but when developing a model or evaluating one, they tend to focus on perceptual similarities between the actual model and what it is designed to represent (AAAS 2007).

• Middle and high school students are more apt to view visual representations such as maps and diagrams as models than they are ideas or abstract models (AAAS 2007).

• Students may lack the notion of the usefulness of a model as being tested against actual observations. They know models can be changed, but at the high school level they may be limited by thinking that a change in a model means adding new information or at the middle school level by thinking that changing a model means replacing a part that was wrong (AAAS 1993).

• Many high school students recognize that models can help them understand the natural world, but they don't believe models can duplicate reality (AAAS 1993).

• Students are more apt to accept the explanatory role of models if many of the material features of the model are similar to the

---

★ Indicates a strong match between the ideas elicited by the probe and a national standard's learning goal.

phenomenon. The more abstract a model is, the less apt students are to recognize its explanatory power (AAAS 1993).

- Middle and high school students tend to think that everything they learn in science is factual. They have difficulty distinguishing between observation and the use of a model to explain a theory (AAAS 1993).

- Students find the idea of multiple models confusing. When multiple models exist, they think that each one represents different aspects of what is being modeled. When multiple models are presented, they tend to think there is one "right one" (AAAS 2007).

Suggestions for Instruction and Assessment

- For younger students, provide opportunities for them to play with toys. Talk about how the toys are like real things only in some ways. They are not the same size, are missing many details, or are not able to do all of the same things.

- Elementary students may need many cycles of reflection and evaluation to overcome their focus on perceptual similarities between the actual model and what it is designed to represent. Have students build a model of a familiar object such as an elbow. If their model looks like but does not act like the real thing, have them talk about this limitation of their model and redesign their model to better represent the function of the item.

- Provide students the opportunity to examine a variety of examples of two- and three-dimensional physical models of the same thing. Analyze and discuss what each model represents and what it doesn't represent. Compare strengths, weaknesses, and limitations of the various models, paying particular attention to scale (relative size, distances), focusing on the proportionality when they are much larger or smaller rather than on the actual number. Propose changes to these models and consider how those changes may better reflect the real thing.

- Provide students the opportunity to examine examples of a physical, conceptual, and mathematical model of the same thing. Compare and contrast what each kind of model can convey and discuss the value of using a variety of kinds of models.

- When models such as atomic/molecular models are used to explain atomic theory, be explicit that the model explains a theory, not an actual observation of atoms and molecules. Point out irrelevant aspects of concrete physical models that can distract from the abstract idea.

- Integrate the unifying theme of models into content domains in a variety of contexts. Use models to examine a variety of real-world phenomena from the living world, physical setting, and technological world.

- Use available online representations (drawings, diagrams, graphs, simulations, and analogies) to help clarify key ideas in science. A collection of reviewed online representations that can be used for science

instruction are found at PRISMS (Phenomena and Representations for Instruction of Science in Middle Schools). To access this collection, go to *http://prisms.mmsa.org*.

- Connect ideas about models to systems thinking. Make models of large systems, such as a watershed, and discuss the boundaries, inputs, outputs, interactions, and consequences as a drop of dye is used to represent a pollutant leaking into a stream.

- Models are important to instruction as a means to make students' thinking visible (Michaels, Shouse, and Schweingruber 2008).

Related NSTA Science Store Publications, NSTA Journal Articles, NSTA SciGuides, NSTA SciPacks, and NSTA Science Objects

Ebert, J. R., N. A. Elliott, and A. Schulz. 2004. Modeling convection. *The Science Teacher* (Sept.): 48–50.

Eichinger, J. 2005. Methods and strategies: Using models effectively. *Science & Children* (Apr.): 43–45.

Finson, K., and J. Beaver. 2007. Time on your hands: Modeling time. *Science Scope* (Jul.): 33–37.

Frazier, R. 2003. Rethinking models. *Science & Children* (Jan.): 29–33.

Gilbert, S. W., and S. W. Ireton. 2003. *Understanding models in Earth and space science.* Arlington, VA: NSTA Press.

Goodwyn, L., and S. Saim. 2007. Modeling mus-

cles. *The Science Teacher* (Dec.): 49–52.

Hitt, A., and J. S. Townsend. 2004. Models that matter. *The Science Teacher* (Mar.): 29–31.

Jones, M. G., M. R. Falvo, A. R. Taylor, and B. P. Broadwell. 2007. *Nanoscale science: Activities for grades 6–12.* Arlington, VA: NSTA Press.

Kane, J. 2004. Geology on a sand budget. *The Science Teacher* (Sept.): 36–39.

Laney, E., and S. Mattox. 2007. Using clay models to understand volcanic mudflows. *Science Scope* (Mar.): 22–25.

Leager, C. R. 2007. Making models. *Science & Children* (Feb.): 50–52.

Littlejohn, P. 2007. Building leaves and an understanding of photosynthesis. *Science Scope* (Apr./May): 22–25.

Michaels, S., A. Shouse, and H. Schweingruber, H. 2008. *Ready, set, science! Putting research to work in K-8 classrooms.* Washington, DC: National Academies Press.

Related Curriculum Topic Study Guide
(Keeley 2005)
"Models"

References

American Association for the Advancement of Science (AAAS). 1993. *Benchmarks for science literacy.* New York: Oxford University Press.

American Association for the Advancement of Science (AAAS). 2007. *Atlas of science literacy.* Vol. 2. (See "Models," pp. 92–93.) Washington, DC: AAAS.

American Association for the Advancement of Science (AAAS). 2008. Benchmarks for science lit-

Is It a System?

Teacher Notes

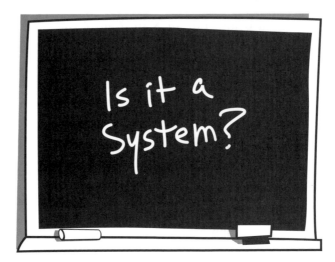

Purpose

The purpose of this assessment probe is to elicit students' ideas about systems. The probe is designed to find out whether students can recognize that things with parts that interact or influence each other are systems.

Related Concept

system

Explanation

The best answer is that everything except for the pile of sand and box of nails can be considered to be a system. If you remove some of the sand, it is still a pile of sand. Removing a sand grain, a cup of sand, or a bucket of sand does not influence whether the sand is still a sandpile, nor do the parts of the sand interact with one another. The nails in the box do not inter-

act with or influence each other. They may be part of a larger system but by themselves they are not a system. You can remove some of the nails and you still have a box of nails. However, there are ways one could stretch this to justify that the sandpile or box of nails is a system. For example, the atoms that make up the sand or nails interact with one another.

Systems range from the simple to the complex. A system is a collection of things (including processes) that have some influence on one another and the whole (AAAS 1989). Systems can be manufactured objects (thermometer, bicycle, cell phone, electrical circuit), life-forms (grasshopper, human body, seed, cell), combinations of living and nonliving things (food web, aquarium, ocean, soil, Earth), physical bodies (volcano, Earth and its Moon), processes (water cycle, hurricane, digestion), or quanti-

tative relationships (Density = Mass/Volume, A + B = C, graph). To be considered a system, the components must interact with or influence each other in some way. Systems are often connected to other systems, may have subsystems, and may be part of larger systems (e.g., human body systems). Their inputs and outputs can include matter, energy, or information.

Curricular and Instructional Considerations

Elementary Students

In the elementary grades, systems begin with parts-and-wholes relationships. Students begin to identify parts of objects such as toy vehicles, animals, dolls, and houses and observe how one part connects to and affects another. This sets the stage for taking apart and reassembling more complicated mechanical systems, emphasizing the importance of the arrangement of parts, and recognizing interactions.

Middle School Students

In the middle grades, students begin thinking from a systems approach, analyzing parts and interactions and identifying subsystems. Disassembly of more complex objects such as clocks or bicycles provides opportunities to describe the interaction of parts, not simply label collections as systems. Projects in which students design, assemble, analyze, and troubleshoot manufactured systems (e.g., battery-powered electrical circuit) and examine biological systems (e.g., organisms in an aquarium) are common at this grade span.

Experiences are provided in which inputs to a system that affect the output are changed (e.g., adding another battery to a circuit, adding more fish to an aquarium). Emphasis is placed on the connections among systems; a battery can be thought of as a system that is also part of a larger system in a circuit, and a fish itself can be considered to be a system that is part of a larger system in an aquarium.

High School Students

Students at the high school level have opportunities to reflect on the value of thinking in terms of systems and to apply the concept in a variety of contexts. Through projects, readings, experiments, and discussion, students analyze the boundaries and components of systems and distinguish the properties of the system from the properties of the parts. Students begin to use the concept of feedback mechanisms (e.g., homeostasis) to explain why things happen and predict changes that may occur.

Administering the Probe

This probe is best used at the upper-elementary, middle, and high school levels. It can be used as a paper and pencil assessment to gather students' ideas for later analysis as well as a stimulus for provoking discussion about systems. It can also be administered as a card sort with small groups sorting examples into two groups—"Is a System" or "Is Not a System"—justifying their reasons as they place each card into a category. Remove items that students may be unfamiliar with and/or add items that connect to systems ideas in your curriculum.

Related Ideas in *National Science Education Standards* (NRC 1996)

K–12 Unifying Concepts and Processes—Systems

- The natural and designed world is too large and complex to comprehend all at once. Scientists define small portions for the convenience of investigation. These portions are referred to as *systems*.

★ A system is an organized group of related objects or components that form a whole. Systems have boundaries, components, resources flow (input and output), and feedback.

- Within systems, interactions between components occur.

- Systems at different levels of organization can manifest different properties and functions.

- Thinking in terms of systems helps keep track of mass, energy, objects, organisms, and events referred to in the other content standards.

- The idea of simple systems encompasses subsystems as well as identifying the structure and function of systems, feedback and equilibrium, and identifying the distinction between open and closed systems.

- Understanding the regularities in systems can develop understanding of basic laws, theories, and models that explain the world.

Related Ideas in *Benchmarks for Science Literacy* (AAAS 1993 and 2008)

Note: Benchmarks revised in 2008 are indicated by (R). New benchmarks added in 2008 are indicated by (N).

K–2 Systems

- Most things are made of parts.

- Something may not work if some of its parts are missing.

- When parts are put together, they can do things that they couldn't do by themselves.

3–5 Systems

- In something that consists of many parts, the parts usually influence one another.

6–8 Systems

★ A system can include processes as well as things.

- Thinking about things as systems means looking for how every part relates to others. The output from one part of a system (which can include material, energy, or information) can become the input to other parts. Such feedback can serve to control what goes on in the system as a whole.

- Any system is usually connected to other systems, both internally and externally. Thus a system may be thought of as containing subsystems and as being a subsystem of a larger system.

- Some portion of the output of a system may be fed back to that system's input. (N)

★ Indicates a strong match between the ideas elicited by the probe and a national standard's learning goal.

- Systems are defined by placing boundaries around collections of interrelated things to make them easier to study. Regardless of where the boundaries are placed, a system still interacts with its surrounding environment. Therefore, when studying a system, it is important to keep track of what enters or leaves the system. (N)

9–12 Systems

- A system usually has some properties that are different from those of its parts but appear because of the interaction of those parts.
- Understanding how things work and designing solutions to problems of almost any kind can be facilitated by systems analysis. In defining a system, it is important to specify its boundaries and subsystems, indicate its relation to other systems, and identify what its input and its output are expected to be.
- Even in some very simple systems, it may not always be possible to predict accurately the result of changing some part or connection.
- The successful operation of a designed system often involves feedback. Such feedback can be used to encourage what is going on in a system, discourage it, or reduce its discrepancy from some desired value. The stability of a system can be greater when it includes appropriate feedback mechanisms. (R)
- Systems may be so closely related that there is no way to draw boundaries that separate all parts of one from all parts of the other. (N)

Related Research

- Elementary students may believe that a system of objects must be doing something in order to be a system or that a system that loses a part of itself is still the same system (AAAS 1993).
- Students of all ages tend to interpret phenomena by noting the qualities of separate objects rather than by seeing the interactions between the parts of a system (e.g., force is considered as a property of bodies rather than as an interaction between bodies; AAAS 1993).
- Students explain changes as a directional chain of cause and effect rather than as two systems interacting (AAAS 1993).

Suggestions for Instruction and Assessment

- Provide students with opportunities to examine a variety of examples of familiar manufactured systems (bicycle, can opener, pencil sharpener, flashlight). Ask questions about what this example of a system does; what the boundaries, inputs, and outputs are; and how the components interact and contribute to the system as a whole. Then ask the same questions about a natural system (e.g., a familiar ecosystem, a cell, the solar system). Emphasize the interactions and influences, not simply the names of the components. A list of questions about systems accompanies the American Association for the Advancement of Science Project 2061 lesson "Seeing the Cell as a System," included in *Resources for Science*

Literacy (AAAS 1997). This lesson can also be accessed online at *www.project2061.org/publications/rsl/online/guide/ch2/hlpsys0.pdf* and *www.project2061.org/publications/rsl/online/guide/ch2/hqsystem.pdf.*

- Challenge students to come up with examples of systems that have the word *system* in them (e.g., solar system, school system, human body systems, ecosystem). Ask them what all of these things have in common that make them systems. Then challenge them to come up with examples of systems that do not include the word *system.*

- Use a FACT (formative assessment classroom technique) such as the Frayer Model, Scientist's Idea strategy, or First Word–Last Word to determine students' ideas about systems before and after instruction (Keeley 2008).

- Integrate the unifying theme of systems into content domains in a variety of contexts. Make explicit connections between systems-thinking and life science (the human body, cells, photosynthesis), Earth/environmental science (ecosystems, climate and weather, Earth system interactions), and physical science (force and motion, energy transfer and transformation) ideas.

- Develop the ideas of input, output, and interactions among components during engineering design exploration and analysis.

- Connect ideas about systems to the idea of models, data collection, and graphing. The purpose of studying systems is to develop the ability to think and analyze in terms of systems. Such thinking can strengthen the skill of identifying regularities and patterns, which supports an understanding of models that explain the world. Prediction from a systems perspective involves using knowledge about the world and an understanding of trends in data to identify and explain observations or changes in advance.

- Challenge students to come up with ideas of things that are not systems. Ask them to apply their lists of characteristics of a system to decide whether the item is a system or not.

Related NSTA Science Store Publications, NSTA Journal Articles, NSTA SciGuides, NSTA SciPacks, and NSTA Science Objects

American Association for the Advancement of Science (AAAS). 2001. *Atlas of science literacy.* Vol. 1. (See "Systems," pp. 132–133.) Washington, DC: AAAS.

Breene, A., and D. Gilewski. 2008. Investigating ecosystems in a biobottle. *Science Scope* (Feb.): 12–15.

Hmelo-Silver, C. E., R. Jordan, L. Liu, S. Gray, M. Demeter, S. Rugaber, S. Vattam, and A. Goel. 2008. Focusing on function: Thinking below the surface of complex natural systems. *Science Scope* (July): 27–35.

Leager, C. R. 2007. Ecosystem in a jar. *Science & Children* (Apr./May): 56–58.

Llewellyn, D., and S. Johnson. 2008. Teaching science through a systems approach. *Science Scope* (July): 21–26.

Ludwig, C., and N. S. Baliga. 2008. Systems concepts effectively taught: Using systems practices. *Science Scope* (July): 16–20.

National Science Teachers Association (NSTA). 2006. Coral ecosystems. NSTA SciGuide. Online at *http://learningcenter.nsta.org/search.aspx?action= quicksearch&text=Coral%20Reef%20Ecosystems*

Related Curriculum Topic Study Guide

(Keeley 2005)

"Systems"

References

American Association for the Advancement of Science (AAAS). 1989. *Science for all Americans.* New York: Oxford University Press.

American Association for the Advancement of Science (AAAS). 1993. *Benchmarks for science literacy.* New York: Oxford University Press.

American Association for the Advancement of Science (AAAS). 1997. *Resources for science literacy.* CD-ROM. New York: Oxford University Press.

American Association for the Advancement of Science (AAAS). 2008. Benchmarks for science literacy online. *www.project2061.org/publications/bsl/online*

Keeley, P. 2005. *Science curriculum topic study: Bridging the gap between standards and practice.* Thousand Oaks, CA: Corwin Press.

Keeley, P. 2008. *Science formative assessment: 75 practical strategies for linking assessment, instruction, and learning.* Thousand Oaks, CA: Corwin Press.

National Research Council (NRC). 1996. *National science education standards.* Washington, DC: National Academy Press.

Life, Earth, and Space Science Assessment Probes

Life, Earth, and Space Science Assessment Probes
Concept Matrix

Core Science Concepts	Life Science							Earth Science				Space Science		
	Is It Food?	Biological Evolution	Chicken Eggs	Adaptation	Is It "Fitter"?	Catching a Cold	Digestive System	Camping Trip	Global Warming	Where Does Oil Come From?	Where Would It Fall?	Moonlight	Lunar Eclipse	Solar Eclipse
Adaptation				✓	✓									
Biological evolution		✓												
Cells							✓							
Climate change									✓					
Common cold						✓								
Conservation of matter			✓											
Digestive system							✓							
Earth history											✓			
Earth's water distribution											✓			
Embryo development			✓											
Food	✓		✓				✓							
Fossil fuel										✓				
Germ theory						✓								
Global warming									✓					
Greenhouse gas									✓					
Heat transfer								✓						
Infectious disease						✓								
Light reflection												✓		
Lunar eclipse													✓	
Moon												✓		
Moon phases												✓		
Natural resource										✓				
Natural selection		✓		✓	✓									
Nonrenewable resource										✓				
Nutrients	✓						✓							
Oceans												✓		
Origin of life		✓												
Solar eclipse														✓
Solar radiation								✓						
Surface of the Earth											✓			
System			✓											
Temperature								✓						
Transformation of matter			✓											
Variation				✓										
Weather								✓						

National Science Teachers Association

Is It Food?

What kinds of things are considered food? Check off the things on the list that are scientifically called food.

___ lettuce   ___ sugar   ___ salt

___ cookies   ___ bread   ___ butter

___ milk   ___ vitamins   ___ water

___ french fries   ___ candy bar   ___ turkey

___ minerals   ___ pancake syrup   ___ banana

___ ketchup   ___ diet soda   ___ flour

Explain your thinking. What definition or "rule" did you use to decide if something can scientifically be called food?

_____

_____

_____

_____

_____

_____

_____

_____

_____

_____

_____

Is It Food?

Teacher Notes

Purpose

The purpose of this assessment probe is to elicit students' ideas about food. The probe is designed to find out if students use a scientific definition of food to distinguish food items from nonfood items.

Related Concepts

food, nutrients

Explanation

The best answers are the following: lettuce, sugar, cookies, bread, butter, milk, french fries, candy bar, turkey, pancake syrup, banana, ketchup, and flour. All of these things are scientifically categorized as food. The scientific definition of food is an organic substance containing carbohydrates, proteins, and/or fats that serves as both fuel and building material

for an organism. This is often confused by our everyday use of the word *food* to mean anything that is ingested.

All foods are nutrients but not all nutrients are food. Nutrients are substances that organisms must take in to carry out their life processes. Nutrients include inorganic and organic materials. Although organisms require inorganic nutrients such as water, vitamins, and salt to grow and survive, they are not a source of energy nor are they used as the building material that primarily makes up the body of an organism. Thus, they are not food from a scientific perspective. Food items contain calories, a measure of the chemical energy from carbohydrate, fat, or protein molecules that make up food. Even though some foods, like lettuce, contain minimal calories, they are still considered food.

Although some foods are considered "good" because they provide a rich source of carbohydrates, proteins, and fats (e.g., peanut butter) and others "bad" because they are of limited nutritional value (e.g., candy, cookies), their ability to provide both energy and building material qualifies them as food. Diet soda contains water and inorganic chemicals and thus is not considered food. Furthermore, the label describes diet soda as having zero calories; thus it does not provide energy. Some materials like flour and butter are not consumed directly as food but are used to make food items. However, they are still considered food in a biological sense.

Curricular and Instructional Considerations

Elementary Students

In the elementary grades, students learn about the needs of organisms, including humans. Through a variety of instructional opportunities, students learn that animals take their food in from the environment by eating plants, animals, or both. They know that all organisms need energy, although energy is still a mysterious concept. Elementary students also learn about food groups and nutrients in the context of human nutrition. They learn the basics of what constitutes good nutrition and how to keep healthy by eating the right foods. This probe is useful in identifying ideas that students have about what makes something "food" before they encounter the more sophisticated ideas about energy flow and matter transformation associated with food.

Middle School Students

In the middle grades, students develop a more sophisticated understanding of food at both a substance and molecular level. Students extend their study of the healthy functioning of the human body and the ways it can be maintained or harmed by diet. This is a time when students can start reading the nutritional labels on food products; paying attention to the carbohydrate, protein, and fat content of foods; learning what a calorie means; and considering what a healthful diet should include.

It is at this time that students can develop a scientific conception of food different from the common, everyday use of the word. Students at this level need to understand that food provides the energy that organisms need to carry out their life processes. They also need to understand that food is broken down into molecules inside the body and that these molecules can be reassembled into the living material of an organism. Students also begin to distinguish substances that are organic in origin from those that are inorganic. This probe is useful in revealing whether students are able to discern the differences between food and other nutrients necessary to carry out life processes.

High School Students

Students at the high school level expand their understanding that food is a source of energy that is released when chemical bonds from carbon-containing molecules are broken during chemical reactions. They are able to distinguish among carbohydrates, fats, and proteins at the molecular level and understand their

roles in metabolic processes. They can differentiate between carbon-containing nutrients and inorganic nutrients and can understand their roles in essential cell functions. This probe is useful at the high school level because it can reveal whether students use a common definition of food and whether they understand what food is from a scientific perspective.

Administering the Probe

Be sure students are familiar with the items on the list. Ask them to cross out any words they are unfamiliar with. Consider having samples of each item to clarify any items students do not recognize. This probe can also be used as a card sort with words or pictures. In small groups, students can sort cards, putting items into three groups—those that represent food, those that do not represent food, and those that they are unsure of—and discussing their reasons for deciding whether each item is considered food. Consider replacing items, or adding additional items, that your students may be more familiar with.

Related Ideas in *National Science Education Standards* (NRC 1996)

· ·

K–4 Characteristics of Organisms

- Organisms have basic needs. For example, animals need air, water, and food; plants require air, water, nutrients, and light.

5–8 Structure and Function in Living Systems

★ Cells carry on the many functions needed to sustain life. This requires that cells take in nutrients, which they use to provide energy for the work they do and to make the materials that a cell or an organism needs.

5–8 Personal Health

★ Food provides energy and nutrients for growth and development.

9–12 The Cell

- Most cell functions involve chemical reactions. Food molecules taken into cells react to provide the chemical constituents needed to synthesize other molecules. Both breakdown and synthesis are made possible by a large set of protein catalysts, called enzymes. The breakdown of some of the food molecules enables the cell to store energy in specific chemicals that are used to carry out the many functions of the cell.

Related Ideas in *Benchmarks for Science Literacy* (AAAS 1993 and 2008)

· ·

Note: Benchmarks revised in 2008 are indicated by (R). New benchmarks added in 2008 are indicated by (N).

K–2 Flow of Matter and Energy

- Both plants and animals need to take in water, and animals need to take in food.

★ Indicates a strong match between the ideas elicited by the probe and a national standard's learning goal.

K–2 Physical Health

- Eating a variety of healthful foods and getting enough exercise and rest help people to stay healthy.

3–5 Flow of Matter and Energy

- Almost all kinds of animals' food can be traced back to plants.

3–5 Physical Health

★ Food provides fuel and materials for growth and repair of body parts. (R)

- Vitamins and minerals, present in small amounts in foods, are essential to keep everything working well.

3–5 Basic Functions

★ From food, people obtain fuel and materials for body repair and growth. (R)

- The indigestible parts of food are eliminated. (R)

6–8 Flow of Matter and Energy

★ Food provides molecules that serve as fuel and building material for all organisms. Organisms that eat plants break down the plant structures to produce the materials and energy the organisms need to survive.

6–8 Basic Functions

- For the body to use food for energy and building materials, the food must first be digested into molecules that are absorbed and transported to cells.

Related Research

- Students consider food as anything useful taken into an organism's body, including water and minerals (Driver et al. 1994).

- Elementary students generally know that there are different foods, that there are good foods and bad foods, and that there are different nutritional outcomes such as variations in size and health. In addition, they are aware of certain limits (drinking just water leads to death; eating only one thing—even one good food—is insufficient for good health). They still may believe, however, that food and water have equivalent nutritional consequences, that the height and weight are similarly influenced by amount of food eaten, and that energy and strength result from exercise but not nutrition. These misconceptions tend to fade by middle school (AAAS 1993).

- Students often give a nonfunctional explanation of the importance of food as something needed to keep animals alive, without noting the role of food in metabolism (Driver et al. 1994).

- Elementary students may know that food is related to growing and being strong and healthy but might not be aware of the physiological mechanisms. By middle school, students know that food undergoes a process of transformation in the body (AAAS 1993).

- Some students may think that food "turns into" energy or that it vanishes after it is eaten. Few elementary students know that food is changed in the stomach and, after

---

★ Indicates a strong match between the ideas elicited by the probe and a national standard's learning goal.

being broken down into other substances, is carried to tissues throughout the body (Driver et al. 1994).

- After 14-year-old students received instruction about the role of food in metabolism, many reverted to their earlier ideas (Driver et al. 1994).

- High school students regard the term *food* to mean essential building materials or an energy source but do not always see that both requirements must be met for a substance to be considered food (Driver et al. 1994).

- Most adults are familiar with the names of dietary components such as protein but not with their functions. Many think vitamins provide most of the energy needs, and food is frequently associated with growth rather than energy (Driver et al. 1994).

Suggestions for Instruction and Assessment

- Take the time to elicit students' definitions of the word *food* and use this as the starting point to build an understanding of the word from a biological perspective. Have students identify the difference between the everyday use of the word and the scientific use. Contrasting the two and providing examples may help them see the difference and begin to use the scientific definition, which restricts the term to substances that serve as building materials and a source of fuel for the organism.

- Expand the concept of food to include examples for all kinds of living organisms, including plants. Plants use sugar just as animals do. Unlike animals, plants are able to use the energy from sunlight to transform inorganic carbon dioxide and water, which they take in from their environment, into sugars. These simple carbohydrates serve as building materials and a source of fuel for the plants and so are considered to be their food.

- As middle or high school students' conception of food develops, introduce, compare, and contrast other commonly used terminology such as *nutrient, carbohydrate, fat, protein, vitamin,* and *mineral*. High school students who have a foundation in organic chemistry can consider these terms from a molecular perspective.

- Have students bring in a variety of "good" and "bad" foods. Use the nutrition labels to discuss their value, paying particular attention to their caloric, carbohydrate, fat, and protein content. High school students can examine the kinds of carbohydrates (simple versus complex), fats (saturated, trans), and protein found in the foods.

- Combine this probe with the probe "Is It Food for Plants?" in Volume 2 of this series (Keeley, Eberle, and Tugel 2007).

Related NSTA Science Store Publications, NSTA Journal Articles, NSTA SciGuides, NSTA SciPacks, and NSTA Science Objects

American Association for the Advancement of Science (AAAS). 2001. *Atlas of science literacy.* Vol. 1. (See "Flow of Energy in Ecosystems" map,

pp. 78–79.) New York: Oxford University Press.

Crowley, J. 2004. Nutritional chemistry. *The Science Teacher* (Apr.): 49–51.

Farenga, S. J., and D. Ness. 2006. Calories, energy, and the food you eat. *Science Scope* (Feb.): 50–52.

Robertson, W. 2006. Science 101: How does the human body turn food into useful energy? *Science & Children* (Mar.): 60–61.

Related Curriculum Topic Study Guide

(Keeley 2005)

"Food and Nutrition"

References

American Association for the Advancement of Science (AAAS). 1993. *Benchmarks for science literacy.* New York: Oxford University Press.

American Association for the Advancement of Science (AAAS). 2007. *Atlas of science literacy.* Vol. 2. (See "Basic Functions," pp. 40–41.) Washington, DC: AAAS.

American Association for the Advancement of Science (AAAS). 2008. Benchmarks for science literacy online. *www.project2061.org/publications/ bsl/online*

Driver, R., A. Squires, P. Rushworth, and V. Wood-Robinson. 1994. *Making sense of secondary science: Research into children's ideas.* London: RoutledgeFalmer.

Keeley, P. 2005. *Science curriculum topic study: Bridging the gap between standards and practice.* Thousand Oaks, CA: Corwin Press.

Keeley, P., F. Eberle, and J. Tugel. 2007. *Uncovering student ideas in science: 25 more formative assessment probes.* Vol. 2. Arlington, VA: NSTA Press.

National Research Council (NRC). 1996. *National science education standards.* Washington, DC: National Academy Press.

Biological Evolution

Four friends were discussing the meaning of the term *biological evolution*. This is what they said:

Mario: "I think it is another term for natural selection."

Sally: "I think it mainly explains how life started."

Cameron: "I think it mainly explains how life changed after it started."

Paz: "I think it includes both how life started and how it changed after it started."

Who do you most agree with? Explain what you think biological evolution is.

_____

_____

_____

_____

_____

_____

_____

Biological Evolution

Teacher Notes

Purpose

The purpose of this assessment probe is to elicit students' ideas about biological evolution. The probe is designed to find out if students distinguish the theory of biological evolution from ideas about the origin of life and the mechanism for biological evolution.

Related Concepts

biological evolution, natural selection, origin of life

Explanation

The best answer is Cameron's: "I think it mainly explains how life changed after it started." It explains that living things share common ancestors. Scientists seek to understand how life started, but this aspect of biology is not the central focus of the theory of

biological evolution. The origin of life is associated with chemical evolution. The study of chemical evolution yields insight into the processes that lead to the generation of the chemical materials essential for the development of life. Regardless of how scientists think life on Earth started, we do know that after life originated it branched and diversified. Natural selection is part of the theory of biological evolution. It is a mechanism that drives evolutionary change in organisms. The theory of biological evolution focuses on explaining life's diversity, and scientists continue to study the relatedness among organisms and how life diversified.

Curricular and Instructional Considerations

Elementary Students

In the elementary grades, the focus is on building a knowledge base about biological diversity to build a foundation for later understanding of the concept of biological evolution. Students learn about life-forms that no longer exist and compare their similarities to present-day organisms. They also examine features of organisms that help the organisms survive in their environments. They examine visible anatomical similarities that help them begin to build evidence for similarities within the diversity of life.

Middle School Students

In middle school, students expand the idea of similarity among seemingly diverse organisms by examining similarities in cells, tissues, and organs as well as similarities in patterns of development and chemical processes such as photosynthesis. This contributes further to building an evidence base for relatedness within the vast diversity of organisms on Earth.

Furthermore, students build a deeper understanding of fossil evidence and Earth's geologic history, solidifying the notion of evolutionary change. They develop an understanding of how successful traits allow individuals to survive and reproduce as well as the effect of environmental changes on organisms and species. Understanding these ideas lays a foundation for understanding the formal concepts of adaptation and natural selection. The formal terminology, *biological evolution* and *natural selection*, is introduced in middle school after students have developed a beginning conceptual understanding of these concepts.

High School Students

Biological evolution is the central theme of modern biology. The foundational ideas and evidence base developed in K–8 can now converge into developing a formal understanding of biological evolution and its mechanism, natural selection. Because of students' readiness to examine molecular evidence and other complexities, combined with their increased skills in examining arguments, high school is the time to develop a clear understanding of biological evolution. In middle school, the emphasis was on selection of individuals with advantageous traits. In high school, the emphasis shifts to include the changing proportions of traits that can result in species changes. Historical perspectives, including Charles Darwin's contribution, provide an opportunity to understand how careful observations lead to solving some of the great puzzles of science.

Administering the Probe

This probe is most appropriate for use at the high school level, although it can be used at the middle school level to ascertain students' preexisting ideas about biological evolution that they may have encountered through the media or other means. Make sure students understand that the probe is focused on biological evolution, not evolution in general.

Related Ideas in *National Science Education Standards* (NRC 1996)

K–4 The Characteristics of Organisms

- Organisms can survive only in environments in which their needs can be met.
- Each plant or animal has different structures that serve different functions in growth, survival, and reproduction.

K–4 Organisms and Their Environments

- When the environment changes, some plants and animals survive and reproduce, and others die or move to new locations.

K–4 Properties of Earth Materials

- Fossils provide evidence about the plants and animals that lived long ago and the nature of the environment at that time.

5–8 Diversity and Adaptations of Organisms

- ★ Biological evolution accounts for the diversity of species developed through gradual processes over many generations. Species acquire many of their unique characteristics through biological adaptation, which involves the selection of naturally occurring variations in populations.
- Millions of species of animals, plants, and microorganisms are alive today. Although different species might look dissimilar, the unity among organisms becomes apparent from an analysis of internal structures, the similarity of their

chemical processes, and the evidence of common ancestry.

9–12 Biological Evolution

- ★ Species evolve over time. Evolution is the consequence of the interactions of (1) the potential for a species to increase its numbers, (2) the genetic variability of offspring due to mutation and recombination of genes, (3) a finite supply of the resources required for life, and (4) the ensuing selection by the environment of those offspring better able to survive and leave offspring.
- ★ The great diversity of organisms is the result of more than 3.5 billion years of evolution that has filled every available niche with life-forms.

Related Ideas in *Benchmarks for Science Literacy* (AAAS 1993 and 2008)

Note: Benchmarks revised in 2008 are indicated by (R). New benchmarks added in 2008 are indicated by (N).

K–2 Evolution of Life

- Some kinds of organisms that once lived on Earth have completely disappeared, although they were something like other organisms that are alive today.

3–5 Evolution of Life

- Individuals of the same kind differ in their characteristics, and sometimes the differences give individuals an advantage in surviving and reproducing.

★ Indicates a strong match between the ideas elicited by the probe and a national standard's learning goal.

- Fossils can be compared with one another and with living organisms according to their similarities and differences. Some organisms that lived long ago are similar to existing organisms, but some are quite different.

6–8 Evolution of Life

- Small differences between parents and off-spring can accumulate (through selective breeding) in successive generations so that descendants are very different from their ancestors.

- Individual organisms with certain traits are more likely than others to survive and have offspring.

- Changes in environmental conditions can affect the survival of individual organisms and entire species.

9–12 Evolution of Life

- ★ The basic idea of biological evolution is that the Earth's present-day species are descended from earlier, distinctly different species. (R)

- Natural selection provides the following mechanism for evolution: Some variation in heritable characteristics exists within every species; some of these characteristics give individuals an advantage over others in surviving and reproducing; and the advantaged offspring, in turn, are more likely than others to survive and reproduce. As a result, the proportion of individuals that have advantageous characteristics will increase. (R)

- Life on Earth is thought to have begun as simple, one-celled organisms about four billion years ago. Once cells with nuclei developed about one billion years ago, increasingly complex multicellular organisms evolved.

- ★ Modern ideas about evolution and heredity provide a scientific explanation for the history of life on Earth as depicted in the fossil record and in the similarities evident within the diversity of existing organisms.

Related Research

- Many people believe that evolution is a theory about the origin of life (University of California Museum of Paleontology 2006).

- Research suggests that students' understanding of evolution is related to their understanding of the nature of science and their general reasoning abilities (AAAS 1993).

Suggestions for Instruction and Assessment

- The word *evolution* is used in many ways in science, including in the terms *biological evolution, chemical evolution, stellar evolution, evolution of the Earth system,* and *evolution of the universe.* Explicitly develop the precise meaning of *biological evolution* so that students distinguish it from other types of change.

- Be aware that some students confuse natural selection with biological evolution. Explicitly develop the notion that natural selection is the mechanism for biological evolution.

★ Indicates a strong match between the ideas elicited by the probe and a national standard's learning goal.

- If students confuse the origin of life with biological evolution, use Charles Darwin's *On the Origin of Species* to compare and contrast the "two origins" by pointing out that "origin of life" deals with the chemical circumstances that produced the first self-replicating molecules that led to the beginning of life, whereas *On the Origin of Species* was a pivotal work that explained species change over time (biological evolution) through natural selection.

- NSTA maintains an extensive website for teaching resources associated with evolution. Go to: *www.nsta.org/publications/ evolution.aspx.*

Related NSTA Science Store Publications, NSTA Journal Articles, NSTA SciGuides, NSTA SciPacks, and NSTA Science Objects

American Association for the Advancement of Science (AAAS). 2001. *Atlas of science literacy.* Vol. 1. (See "Biological Evolution" map, pp. 80–81.) Washington, DC: AAAS.

Biological Sciences Curriculum Study (BSCS). 2005. *The nature of science and the study of biological evolution.* Colorado Springs, CO: BSCS.

Bybee, R., ed. 2004. *Evolution in perspective: The science teacher's compendium.* Arlington, VA: NSTA Press.

Diamond, J., C. Zimmer, E. M. Evans, L. Allison, and S. Disbrow, eds. 2006. *Virus and the whale: Exploring evolution in creatures large and small.* Arlington, VA: NSTA Press.

Kampourakis, K. 2006. The finche's beak: Introducing evolutionary concepts. *Science Scope* (Mar.): 14–17.

McComas, W. 2008. *Investigating evolutionary biology in the laboratory.* Dubuque, IO: Kendall Hunt.

National Academy of Sciences. 2008. *Science, evolution, and creationism.* Washington, DC: National Academies Press.

Scotchmoor, J., and A. Janulaw. 2005. Understanding evolution. *The Science Teacher* (Dec.): 28–29.

Related Curriculum Topic Study Guides

(Keeley 2005)

"Biological Evolution"

"Origin of Life"

References

American Association for the Advancement of Science (AAAS). 1993. *Benchmarks for science literacy.* New York: Oxford University Press.

American Association for the Advancement of Science (AAAS). 2008. Benchmarks for science literacy online. *www.project2061.org/publications/ bsl/online*

Keeley, P. 2005. *Science curriculum topic study: Bridging the gap between standards and practice.* Thousand Oaks, CA: Corwin Press.

National Research Council (NRC). 1996. *National science education standards.* Washington, DC: National Academy Press.

University of California Museum of Paleontology. 2006. Understanding evolution. *http://evolution. berkeley.edu*

Chicken Eggs

The students in Mrs. Bartoli's class were studying how chickens develop from an egg. The students put a dozen freshly laid, fertilized chicken eggs in an incubator. They wondered what would happen to the mass of an egg as the chick inside developed. This is what the students thought:

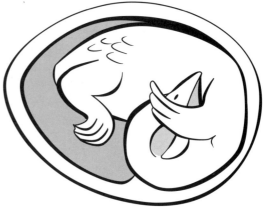

Group A: "We think an egg will gain mass. An egg's mass is more just before hatching than when the egg was laid."

Group B: "We think an egg will lose mass. An egg's mass is less just before hatching than when the egg was laid."

Group C: "We think the mass of an egg stays the same as the chick develops inside."

Which group do you most agree with? Explain your thinking.

_____

_____

_____

_____

_____

_____

_____

_____

_____

Chicken Eggs

Teacher Notes

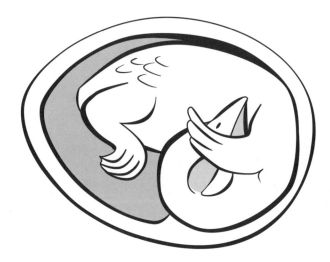

Purpose

The purpose of this assessment probe is to elicit students' ideas about food, transformation of matter, growth and development, conservation of mass, and systems. The concepts underlying this probe are complex. It is not important that students know exactly what happens to the mass of an egg and why. Rather, this probe is used as an interesting context to draw out their ideas about several interrelated concepts in science.

Related Concepts

conservation of matter, embryo development, food, system, transformation of matter

Explanation

The best answer is Group B's: "We think an egg will lose mass. An egg's mass is less just

before hatching than when the egg was laid." During normal incubation, chicken eggs lose approximately 16%–18% of their original mass (Snyder and Birchard 2005). On an average, bird eggs from small hummingbirds to large ostriches lose 15% of their original weight. This weight loss is primarily the result of water vapor passing through the permeable eggshell. Water vapor is a waste product of metabolism.

The egg yolk serves as food for the developing embryo inside. This food is used for the energy the embryo needs to carry out life processes, such as respiration. During respiration, carbon dioxide and water vapor are released as waste products. Molecules from the food are converted into the building material the embryo needs for growth and development.

Intuitively it would seem that the egg would weigh more after the chick has devel-

oped inside. The liquid matter (yolk and "white part") inside the freshly laid egg is transformed into the body tissues of the embryo, which continues to grow and develop as cells divide. The yolk provides the energy and source of building material the chick needs for its development. Although oxygen does diffuse through the porous cell and is used during respiration, most of the material the embryo needs for development is packaged inside the cell at the time it is laid. Because the eggshell is permeable to gases, oxygen enters into the egg through the shell and some water vapor diffuses through the eggshell to the outside environment. If the eggshell were a perfect closed system, there would be no change in mass.

It is not important that students know that an egg can lose a significant percentage of its original mass. What is significant is that they recognize that matter is conserved during the transformation of the egg material and development of the chick but that some of this matter may escape through the egg because it is not a closed system.

Curricular and Instructional Considerations

Elementary Students

In the elementary grades, students study the life cycles of different organisms. Incubating chicken eggs is a common activity in some classrooms. Students at this level can examine eggs, learning that developing embryos, like all animals, need food and that the yolk is the source of the embryo's food. However, the more complex notion of food being transformed into the body material of the embryo should wait until middle school.

Middle School Students

In the middle grades, students develop a scientific conception of what food is and how it provides energy as well as building material for organisms. At this level, students can begin to understand the transformation of the yolk into the body material of the developing chick as a result of chemical reactions and cell division. They also develop the idea of open versus closed systems and can use this idea to consider whether some materials can diffuse in and out of living and nonliving membranes or other porous materials.

High School Students

At this level, students investigate more complex ideas about embryology, chemical processes of metabolism, and passage of materials in and out of an open system. They should know about the breakdown and recombination of molecules during biochemical changes and how matter is conserved in each of these changes. They should recognize the semipermeable nature of membranes and porosity of other seemingly impermeable materials and explain how and why materials pass from an internal to external environment or vice versa.

Administering the Probe

This probe can be accompanied by a visual representation that shows the changes inside a chicken egg at different stages of development.

However, be aware that the representation may reinforce the misconception that the egg would weigh more because it looks like there is more "stuff" inside the egg with the late stage embryo compared with the freshly laid egg that contains mostly liquid-like matter. When used as a discussion prompt, this probe can lead to a very lively discussion and argument among students that elicits ideas about food, conservation of matter, transformations, role of gases, and open and closed systems.

Related Ideas in *National Science Education Standards* (NRC 1996)

. . . .

K–4 The Characteristics of Organisms

- Organisms have basic needs—for example, animals need air, water, and food.

K–4 Life Cycles of Organisms

- Plants and animals have life cycles that include being born, developing into adults, reproducing, and eventually dying.

5–8 Structure and Function in Living Systems

- Cells carry on the many functions needed to sustain life. They grow and divide, thereby producing more cells.

5–8 Properties and Changes of Properties in Matter

- In chemical reactions, the total mass is conserved.

5–8 Regulation and Behavior

- All organisms must be able to obtain and use resources, grow, reproduce, and maintain stable internal conditions while living in a constantly changing environment.

9–12 Cells

- Every cell is surrounded by a membrane that separates it from the outside world.

- ★ Most cell functions involve chemical reactions. Food molecules taken into cells react to provide the chemical constituents needed to synthesize other molecules.

- In the development of multicellular organisms, the progeny from a single cell form an embryo in which the cells multiply and differentiate to form the many specialized cells, tissues, and organs that make up the final organism.

9–12 Matter, Energy, and Organization in Living Systems

- ★ As matter and energy flow through different levels of organization of living systems—cells, organs, organisms, and communities—and between living systems and the physical environment, chemical elements are recombined in different ways. Each recombination results in storage and dissipation of energy into the environment as heat. Matter and energy are conserved in each change.

★ Indicates a strong match between the ideas elicited by the probe and a national standard's learning goal.

Related Ideas in *Benchmarks for Science Literacy* (AAAS 1993 and 2008)

K–2 Flow of Matter and Energy

- Plants and animals both need to take in water, and animals need to take in food.

3–5 Flow of Matter and Energy

- Some source of "energy" is needed for all organisms to stay alive and grow.

6–8 Structure of Matter

★ No matter how substances within a closed system interact with one another or how they combine or break apart, the total mass of the system remains the same. The idea of atoms explains the conservation of matter: If the number of atoms stays the same no matter how the same atoms are rearranged, then their total mass stays the same.

6–8 Cells

- Cells repeatedly divide to make more cells for growth and repair.
- About two-thirds of the weight of cells is accounted for by water, which gives cells many of their properties.

6–8 Flow of Matter and Energy

★ Food provides molecules that serve as fuel and building material for all organisms.

9–12 Cells

- Every cell is covered by a membrane that controls what can enter and leave the cell.

Related Research

- Field tests of this probe with middle and high school students reveal that most students think the mass will be greater because the chick inside the egg is getting bigger. Some students use conservation reasoning to say the mass stays the same. Although their ideas about conservation of mass during the transformation of matter are correct, they fail to consider whether the egg is a closed system.

- Some students use an intuitive rule of "more A, more B" (Stavy and Tirosh 2000). They reason that if the chick gets bigger inside the egg, then the mass or weight of the egg increases.

- Children often do not recognize that food is the material basis for growth—that is, that the food becomes transformed and incorporated into the body, thus making the body bigger (Driver et al. 1994).

- In a study conducted by Russell and Watt (1989), elementary children assumed that the growth inside an egg is associated with an increase in mass within what they assumed was a closed system. They described the process of growth as creating new material rather than transforming material (yolk) that was in the egg. Only a very small minority considered some type of transformation of the contents inside the egg into a complete chick.

- Some ideas about embryonic development contribute to students' notions about what is happening inside the egg. Some children believe that the chick had always been there

---

★ Indicates a strong match between the ideas elicited by the probe and a national standard's learning goal.

inside the egg waiting until it was time to hatch. Others thought all the parts of the chick were there when the egg was laid and that they came together in the egg (Driver et al. 1994).

Suggestions for Instruction and Assessment

- This probe can be used as a P-E-O probe (Predict, Explain, Observe) with activities that involve egg incubation. Have students predict what would happen to the mass of the eggs as the eggs develop over the course of their incubation. Have students explain the reasons for their predictions. Then have students test their ideas and observe the decrease in mass. Encourage students to come up with alternative explanations to account for the discrepancy between their predictions and their observations.

- Probe further to find out students' conceptions of an open versus closed system in relation to the egg. One way to show that water can pass through an egg shell is to place a raw egg in corn syrup. The mass of the egg in corn syrup will decrease because water from inside the egg flows through the membrane and shell into the syrup. It moves from a higher concentration inside the egg to a lower concentration in the corn syrup. The corn syrup molecules are too large to pass into the egg. You can also try this with molasses. Careful observations will reveal a thin layer of water resting on top of the molasses.

- Find the mass of a raw egg and leave it in a warm, dry area for two to three weeks. Find the mass of the egg again, noticing a decrease in mass. Encourage students to propose ideas about where the loss in mass came from. Did anything leave the egg?

- *Safety caution:* Always have students wash their hands after handling eggs.

- Use a graphic of a chick embryological development such as the one at *http://msucares.com/poultry/reproductions/poultry_chicks_embryo.html* (Mississippi State University Extension Service 2004). Have students propose ideas about (a) where the material is coming from that leads to the growth of the chick inside the egg and (b) the chemical life processes that are occurring.

Related NSTA Science Store Publications, NSTA Journal Articles, NSTA SciGuides, NSTA SciPacks, and NSTA Science Objects

American Association for the Advancement of Science (AAAS). 2001. *Atlas of science literacy.* Vol. 1. (See "Flow of Matter in Ecosystems" map, pp. 76–78.) Washington, DC: AAAS.

Related Curriculum Topic Study Guides

(Keeley 2005)

"Reproduction, Growth, and Development (Life Cycles)"

"Food and Nutrition"

"Conservation of Matter"

"Systems"

References

American Association for the Advancement of Science (AAAS). 1993. *Benchmarks for science literacy.* New York: Oxford University Press.

American Association for the Advancement of Science (AAAS). 2008. Benchmarks for science literacy online. *www.project2061.org/publications/bsl/online*

Driver, R., A. Squires, P. Rushworth, and V. Wood-Robinson. 1994. *Making sense of secondary science: Research into children's ideas.* London: RoutledgeFalmer.

Keeley, P. 2005. *Science curriculum topic study: Bridging the gap between standards and practice.* Thousand Oaks, CA: Corwin Press.

Mississippi State University Extension Service. 2004. Poultry: Stages in chick embryo development. *http://msucares.com/poultry/reproductions/poultry_chicks_embryo.html*

National Research Council (NRC). 1996. *National science education standards.* Washington, DC: National Academy Press.

Russell, T., and D. Watt, eds. 1990. *Growth.* Primary SPACE Project Research Report. Liverpool, UK: Liverpool University Press.

Snyder, G., and G. Birchard. 1982. Water loss and survival in embryos of the domestic chicken. *Journal of Experimental Zoology* 219(1): 115–117.

Stavy, R., and D. Tirosh. 2000. *How students (mis-) understand science and mathematics: Intuitive rules.* New York: Teachers College Press.

Adaptation

Three friends were arguing about what would happen if a population of rabbits from a warm, southern climate were moved to a cold, northern climate. This is what they said:

Bernie: "I think all of the rabbits will try to adapt to the change."

Leo: "I think most of the rabbits will try to adapt to the change."

Phoebe: "I think few or none of the rabbits will try to adapt to the change."

Which person do you most agree with and why? Explain your ideas about adaptation.

_____

_____

_____

_____

_____

_____

_____

_____

_____

_____

_____

Adaptation

Teacher Notes

Purpose

The purpose of this assessment probe is to elicit students' ideas about biological adaptation. The probe is designed to find out if students think animals intentionally adapt to a change in their environment.

Related Concepts

adaptation, natural selection, variation

Explanation

The best answer is Phoebe's: "I think few or none of the rabbits will try to adapt to the change." The key word here is *try*. Biological adaptation involves genetic variation that allows some individuals to survive a particular change, such as a change in the environment, better than others. These individuals are then able to survive and reproduce, passing on their genes to succes-

sive generations of offspring that will be better adapted for the particular environment. This process is called natural selection, and it leads to adaptation. If the genetic variation that allows an individual to survive the change is not present, the individual cannot intentionally change its structure, physiology, or behavior in an attempt to "try" to adapt to the change and pass on its genes so that its offspring will be adapted. Either the genes are there that allow the rabbit to survive and pass on its traits that enhance survival to its offspring (natural selection) or they are not there. If they are not there, the rabbits can't intentionally adapt or change their genes by "trying." Adaptation is not intentional. The rabbits may try to survive by acclimating to the change, but trying to survive is different in a biological sense from trying to adapt. Another problem is the common use of the verb *adapt*,

which implies that an action is being taken by an individual.

Curricular and Instructional Considerations

Elementary Students

In the elementary grades, students build understandings of biological concepts through direct experience with living things and their habitats. They observe and learn about structures, functions, and behaviors that help organisms survive in their environments. They develop an understanding that some organisms are better suited than others to survive in certain environments. They develop beginning ideas about heredity—that is, that some characteristics are inherited and passed on to offspring. These basic ideas establish a foundation that will lead to a later understanding of natural selection.

Middle School Students

Understanding adaptation is still particularly troublesome at this level. Many students think *adaptation* means that individuals change in deliberate ways in response to changes in the environment (NRC 1996). At this level, it is important to develop the idea of variations in populations of organisms that may give some individuals an advantage in surviving, reproducing, and passing on those traits to their offspring. Teaching students about the selection of individuals is the first step in helping them understand natural selection as a mechanism for species' change.

High School Students

Biological evolution and its mechanism, natural selection, are major focuses of high school biology. At the high school level, students shift from a focus on selection of individuals with certain traits that help them survive to a focus on the changing proportion of such traits in a population of organisms. Their growing understanding of genetics builds on middle school ideas about variation. However, students at this level may still hold on to the misconception that adaptations can be controlled by an individual.

Administering the Probe

Make sure students understand the probe is a hypothetical situation and that there is a drastic change in the environment when the rabbits move from the southern climate to the northern climate. This change also involves more than just temperature. There may be changes in food, shelter, and predators as well. You might consider having students describe each of the environments first. Feel free to change the context of the probe to an animal and two different environments with which your students are most familiar.

Related Ideas in *National Science Education Standards* (NRC 1996)

K–4 The Characteristics of Organisms

- Organisms have basic needs. Organisms can survive only in environments in which their needs can be met.

- Each plant or animal has different structures that serve different functions in growth, survival, and reproduction.

K–4 Organisms and Their Environments

★ An organism's patterns of behavior are related to the nature of that organism's environment, including the kinds and numbers of other organisms present, the availability of food and resources, and the physical characteristics of the environment. When the environment changes, some plants and animals survive and reproduce, and others die or move to new locations.

5–8 Regulation and Behavior

- All organisms must be able to obtain and use resources, to grow, to reproduce, and to maintain stable internal conditions while living in a constantly changing environment.

5–8 Diversity and Adaptations of Organisms

★ Species acquire many of their unique characteristics through biological adaptation, which involves the selection of naturally occurring variations in populations. Biological adaptations include changes in structures, behaviors, or physiology that enhance survival and reproductive success in a particular environment.

9–12 Biological Evolution

- Species evolve over time. Evolution is the consequence of the interactions of (1) the potential for a species to increase its numbers, (2) the genetic variability of offspring due to mutation and recombination of genes, (3) a finite supply of the resources required for life, and (4) the ensuing selection by the environment of those offspring better able to survive and leave offspring.

9–12 The Behavior of Organisms

★ Like other aspects of an organism's biology, behaviors have evolved through natural selection. Behaviors often have an adaptive logic when viewed in terms of evolutionary principles.

Related Ideas in *Benchmarks for Science Literacy* (AAAS 1993 and 2008)

Note: Benchmarks revised in 2008 are indicated by (R). New benchmarks added in 2008 are indicated by (N).

K–2 Heredity

- There is variation among individuals of one kind within a population.

K–2 Evolution of Life

- Different plants and animals have external features that help them thrive in different kinds of places.

3–5 Interdependence of Life

- For any particular environment, some kinds of plants and animals thrive, some do not live as well, and some do not survive at all. (R)

★ Indicates a strong match between the ideas elicited by the probe and a national standard's learning goal.

3–5 Evolution of Life

★ Individuals of the same kind differ in their characteristics, and sometimes the differences give individuals an advantage in surviving and reproducing.

6–8 Evolution of Life

★ Individual organisms with certain traits are more likely than others to survive and have offspring.

★ Changes in environmental conditions can affect the survival of individual organisms and entire species.

9–12 Heredity

• Some new gene combinations make little difference, some can produce organisms with new and perhaps enhanced capabilities, and some can be harmful.

9–12 Evolution of Life

• Natural selection provides the following mechanism for evolution: Some variation in heritable characteristics exists within every species; some of these characteristics give individuals an advantage over others in surviving and reproducing; and the advantaged offspring, in turn, are more likely than others to survive and reproduce. As a result, the proportion of individuals that have advantageous characteristics will increase. (R)

• Heritable characteristics can be observed at molecular and whole-organism levels—in structure, chemistry, or behavior.

Related Research

• Many students tend to see adaptation as an intention by the organism to satisfy a desire or need for survival (Driver et al. 1994).

• Middle and high school students may believe that organisms are able to intentionally change their bodily structure to be able to live in a particular habitat or that organisms respond to a changed environment by seeking a more favorable environment. It has been suggested that the language about adaptation used by teachers or textbooks may cause or reinforce these beliefs (AAAS 1993, p. 342).

• Many students ages 12–16 display Lamarckian beliefs about inheritance of acquired characteristics. This belief has been demonstrated both before and after instruction in genetics and evolution (Driver et al. 1994).

Suggestions for Instruction and Assessment

• Refrain from using the words *adapt* (particularly as an action verb that implies intentionality) or *adaptation* in elementary grades. Instead talk about characteristics and features that help organisms live in their environments. Teachers' early use of the words *adapt* or *adaptation*, before students understand the genetic basis for passing on traits that enable an organism to adapt, may imply that plants and animals intentionally adapt. This idea is particularly resistant to change at the middle level, perhaps because the idea of intentional adaptation was developed early on.

★ Indicates a strong match between the ideas elicited by the probe and a national standard's learning goal.

- Some individual organisms, such as humans, do control changes in their structure or behavior in response to an environmental change and are said to "adapt." *Acclimatize* would be a better word to use for these noninheritable changes made by an individual during its lifetime.

- A common activity used in elementary and middle school is to have students design an imaginary organism that is "adapted" to a particular habitat. Be aware, though, that this activity may reinforce the idea that an individual intentionally adapts during its lifetime, rather than species adaptation.

- Compare and contrast with students the everyday common use of the word *adaptation* with the scientific meaning of the word. Add this to students' growing number of examples of the way we use words in our society that are not always the same as the way they are used in science.

- Help students develop the notion that natural selection leads to adaptation and not the other way around (that adaptation leads to natural selection). It is the genetic variation that leads to natural selection and thus to adaptation of a population.

Related NSTA Science Store Publications, NSTA Journal Articles, NSTA SciGuides, NSTA SciPacks, and NSTA Science Objects

American Association for the Advancement of Science (AAAS). 2001. *Atlas of science literacy.* Vol. 1. (See "Natural Selection" map, pp. 82–83). Washington, DC: AAAS.

Biological Sciences Curriculum Study (BSCS). 2005.

The nature of science and the study of biological evolution. Colorado Springs, CO: BSCS.

Kampourakis, K. 2006. The finches beak: Introducing evolutionary concepts. *Science Scope* (Mar.): 14–17.

Scotchmoor, J., and A. Janulaw. 2005. Understanding evolution. *The Science Teacher* (Dec.): 28–29.

Stebbins, R., D. Ipsen, G. Gillfillan, J. Diamond, and J. Scotchmoor. Rev. ed. 2008. *Animal coloration: Activities on the evolution of concealment* Rev. ed. Arlington, VA: NSTA Press.

Related Curriculum Topic Study Guides
(Keeley 2005)
"Adaptation"
"Natural and Artificial Selection"
"Variation"

References

American Association for the Advancement of Science (AAAS). 1993. *Benchmarks for science literacy.* New York: Oxford University Press.

American Association for the Advancement of Science (AAAS). 2008. Benchmarks for science literacy online. *www.project2061.org/publications/bsl/online*

Driver, R., A. Squires, P. Rushworth, and V. Wood-Robinson. 1994. *Making sense of secondary science: Research into children's ideas.* London: RoutledgeFalmer.

Keeley, P. 2005. *Science curriculum topic study: Bridging the gap between standards and practice.* Thousand Oaks, CA: Corwin Press.

National Research Council (NRC). 1996. *National science education standards.* Washington, DC: National Academy Press.

Is It "Fitter"?

Natural selection is sometimes described as "survival of the fittest." Four friends were arguing about what the phrase "survival of the fittest" means. This is what they said:

Dora: "I think 'fit' means bigger and stronger."

Lance: "I think 'fit' means more apt to reproduce."

Felix: "I think 'fit' means able to run faster."

Hap: "I think 'fit' means more intelligent."

Which person do you most agree with? Explain what you think "survival of the fittest" means.

_____

_____

_____

_____

_____

_____

_____

_____

Is It "Fitter"?

Teacher Notes

Purpose

The purpose of this assessment probe is to elicit students' ideas about natural selection. The probe is designed to find out how students interpret the commonly used phrase "survival of the fittest."

Related Concepts

adaptation, natural selection, variation

Explanation

The best answer is Lance's: "I think 'fit' means more apt to reproduce." Size, swiftness, strength, and intelligence are key factors often related to survival, but they do not always determine whether an organism is most "fit" to reproduce and pass its genes on to offspring. The key idea portrayed in the phrase *survival of the fittest* is that *fittest* means best suited to

survive and reproduce. In some instances, size is an advantage. For example, a large male sea lion is more apt to be the dominant male who breeds with a harem of female sea lions. Because of his impressive size, the females are more attracted to him and the other males are intimidated. Because the other males fear him, he has less competition for food as well. Therefore, he is more apt to survive and reproduce because he can outcompete the other males for food and mates. On the other hand, imagine breeding two dogs. One dog is a very large, muscular, big-boned, *fit* male. This male is bred to the same type of dog, a petite female. However, the puppies are too large and cannot be delivered without a caesarean section. In this case, the bigger male wasn't necessarily the *fittest*. If this had happened in nature, the female dog probably would have died in birth

and the male would have lost his chance to pass his genes on to offspring. Examples also exist in the plant world. Flower breeders have developed large dahlias that are prized for their huge flowers. However, these large dahlias must be supported with wires and stakes as they grow because they would fall to the ground and possibly die without support (BSCS 2005). Intelligence, swiftness in escaping danger, and strength are also factors that can contribute to survival, yet they do not always equate with reproductive success.

Curricular and Instructional Considerations

. . . .

Elementary Students

In the elementary grades, students build understandings of biological concepts through direct experience with living things and their habitats. They observe and learn about structures, functions, and behaviors that help organisms survive in their environments. They develop an understanding that some organisms are better suited than others to survive in certain environments. They develop precursor ideas to the concept of natural selection, such as the idea that some characteristics are inherited and passed on to offspring. They may have heard the phrase *survival of the fittest* and equate *fit* with strength and size.

Middle School Students

Students at the middle level formally develop an understanding of the concept of natural selection. Often the phrase *survival of the fittest*

is used without defining what is meant by *fit*, thus leading to students' erroneous misinterpretation of the intent of this phrase. At this level, it is important to develop the idea of variations in populations of organisms that may give some individuals an advantage in surviving, reproducing, and passing on those traits to their offspring. Teaching students about the selection of individuals is the first step in helping them understand natural selection as a mechanism for species' change.

High School Students

Biological evolution and its mechanism, natural selection, are major focuses of high school biology. At the high school level, students shift from a focus on selection of individuals with certain traits that help them survive to a focus on the changing proportion of such traits in a population of organisms. Their growing understanding of genetics builds on middle school ideas about variation. However, students at this level may still hold on to a misconception that "survival of the fittest" means that "physically fit" species are more *fit* than smaller ones.

Administering the Probe

This probe is best used at the middle or high school level. Make sure students have encountered the concept of natural selection before using this probe as is. If they have not encountered this term, consider removing the first sentence from the prompt. This probe is best used to engage students in argumentation about the phrase *survival of the fittest* and what it means.

Related Ideas in *National Science Education Standards* (NRC 1996)

K–4 The Characteristics of Organisms

- Each plant or animal has different structures that serve different functions in growth, survival, and reproduction.

5–8 Regulation and Behavior

★ All organisms must be able to obtain and use resources, to grow, to reproduce, and to maintain stable internal conditions while living in a constantly changing environment.

5–8 Diversity and Adaptations of Organisms

★ Species acquire many of their unique characteristics through biological adaptation, which involves the selection of naturally occurring variations in populations. Biological adaptations include changes in structures, behaviors, or physiology that enhance survival and reproductive success in a particular environment.

9–12 Biological Evolution

- Species evolve over time. Evolution is the consequence of the interactions of (1) the potential for a species to increase its numbers, (2) the genetic variability of offspring due to mutation and recombination of genes, (3) a finite supply of the resources required for life, and (4) the ensuing selection by the environment of those offspring better able to survive and leave offspring.

Related Ideas in *Benchmarks for Science Literacy* (AAAS 1993 and 2008)

Note: Benchmarks revised in 2008 are indicated by (R). New benchmarks added in 2008 are indicated by (N).

K–2 Heredity

- There is variation among individuals of one kind within a population.

3–5 Evolution of Life

★ Individuals of the same kind differ in their characteristics, and sometimes the differences give individuals an advantage in surviving and reproducing.

6–8 Evolution of Life

★ Individual organisms with certain traits are more likely than others to survive and have offspring.

9–12 Heredity

- Some new gene combinations make little difference, some can produce organisms with new and perhaps enhanced capabilities, and some can be deleterious.

9–12 Evolution of Life

★ Natural selection provides the following mechanism for evolution: Some variation in heritable characteristics exists within every species; some of these characteristics give individuals an advantage over others in surviving and reproducing; and the advantaged offspring, in turn, are more

---

★ Indicates a strong match between the ideas elicited by the probe and a national standard's learning goal.

likely than others to survive and reproduce. As a result, the proportion of individuals that have advantageous characteristics will increase. (R)

Related Research

• Field-test results of this probe showed that many middle school students are more apt to think of "fit" as big and strong. They relate "fittest" to physical fitness, rather than a species' ability to survive and reproduce.

• Many people mistakenly think that the biggest individual is the fittest (BSCS 2005).

Suggestions for Instruction and Assessment

• Have students generate examples of when "bigger is better" in relation to an organism's ability to survive and reproduce. Then generate examples in which it may be a detriment. Try this with other characteristics to show that there is not one characteristic that always determines *fitness for survival.*

• Compare and contrast with students the common use of the word *fit* (as in physical fitness) with the scientific meaning of the word in relation to natural selection. Add this to students' growing number of examples of the way we use words in our society that are not always the same as the way they are used in science.

Related NSTA Science Store Publications, NSTA Journal Articles, NSTA SciGuides, NSTA SciPacks, and NSTA Science Objects

American Association for the Advancement of Science (AAAS). 2001. *Atlas of science literacy.* Vol. 1. (See "Natural Selection" map, pp. 82–83.) Washington, DC: AAAS.

Benz, R. 2000. *Ecology and evolution: Islands of change.* Arlington, VA: NSTA Press.

Biological Sciences Curriculum Study (BSCS). 2005. *The nature of science and the study of biological evolution.* Colorado Springs, CO: BSCS.

Diamond, J., C. Zimmer, E. M. Evans, L. Allison, and S. Disbrow, eds. 2006. *Virus and the whale: Exploring evolution in creatures large and small.* Arlington, VA: NSTA Press.

Kampourakis, K. 2006. The finches' beaks: Introducing evolutionary concepts. *Science Scope* (Mar.): 14–17.

Scotchmoor, J., and A. Janulaw. 2005. Understanding evolution. *The Science Teacher* (Dec.): 28–29.

Related Curriculum Topic Study Guide

(Keeley 2005)

"Natural and Artificial Selection"

References

American Association for the Advancement of Science (AAAS). 1993. *Benchmarks for science literacy.* New York: Oxford University Press.

American Association for the Advancement of Sci-

ence (AAAS). 2008. Benchmarks for science literacy online. *www.project2061.org/publications/bsl/online*

Biological Sciences Curriculum Study (BSCS). 2005. *The nature of science and the study of biological evolution.* Colorado Springs, CO: BSCS.

Keeley, P. 2005. *Science curriculum topic study: Bridging the gap between standards and practice.* Thousand Oaks, CA: Corwin Press.

National Research Council (NRC). 1996. *National science education standards.* Washington, DC: National Academy Press.

Catching a Cold

Have you ever been sick with a cold? People have different ideas about what causes a cold. Check off the things that cause you to "catch a cold."

____ having a fever

____ being wet

____ being wet and cold

____ germs

____ spoiled food

____ not getting enough sleep

____ lack of exercise

____ cold weather

____ dry air

____ imbalance of body fluids

Explain your thinking. Describe how people "catch a cold."

_____

_____

_____

_____

_____

_____

_____

Catching a Cold

Teacher Notes

Purpose

The purpose of this assessment probe is to elicit students' ideas about infectious disease. The probe is designed to find out whether students use the germ theory to explain what causes an infectious disease like the common cold.

Related Concepts

common cold, germ theory, infectious disease

Explanation

The best answer is "germs." The common cold is an infectious disease caused by a virus and transmitted between two people—one who is contagious and one who picks up the contagion (virus). The *cause* is the virus (germs), *transmission* is how it is spread, and other factors contribute to a weakened immune system that is less effective in fighting off the virus

in the human body. The virus is transmitted through respiratory secretions. The virus can be picked up by breathing in the virus when it is spread in an aerosol form generated by the sick person's coughing or sneezing. It can also be picked up from direct contact with saliva or nasal secretions containing the virus as well as indirectly from surfaces that have been contaminated by a person's saliva, respiratory aerosols, or nasal secretions. This is why hand washing is so important. Most cold germs are picked up by touching contaminated surfaces and transferring the virus from an object to the mouth. In all of these cases of transmission, what causes the cold is the virus.

A fever is a physiological response to the virus, not a cause. Feeling cold and chilled, being wet, being wet and cold, and not getting enough sleep or exercise are all factors that can

contribute to a weakened immune system that is less effective in fighting off the virus as it multiplies inside the body's cells. These factors that lower resistance are not the actual cause of a cold. For example, one does not catch a cold merely by being wet and cold. A virus must enter the body in order to cause a cold. Food spoils as a result of bacterial growth and results in a bacterial infection that causes gastrointestinal problems, not a common cold.

Although colds occur more often in the winter months, the cold weather itself does not cause the common cold. During cold weather months, people spend more time inside in close proximity to each other, thus spreading the virus more easily. Also the hot, dry air that results from heating during the wintertime dries out the mucus membranes of the throat and nose and makes them less effective barriers against infection by the common cold virus.

Curricular and Instructional Considerations

Elementary Students

In the elementary grades, students should have a variety of experiences that provide initial understandings of various science-related personal and societal health challenges (NRC 1996). Children at this age use the word *germs* for all microbes, as they may not yet be ready to distinguish between bacteria and viruses. They develop an understanding of good health factors, such as nutrition, exercise, keeping warm and dry, and sleep, but they have difficulty distinguishing between the factors that promote good health in general and the causes of infectious diseases. At this stage they should be taught how communicable diseases such as colds are transmitted, and the reason for hand washing should be explained, reinforced, and practiced in school and at home. Later in the elementary grades, students begin to learn about some of the body's defense mechanisms that prevent or overcome infectious diseases such as colds.

Middle School Students

In the middle grades, students build upon their K–4 understandings of health and disease to recognize the role of microorganisms in causing illness. This is a good time to introduce the germ theory of diseases from the historical perspective of Louis Pasteur's discovery and to discuss how technology (microscopes) has made germs visible.

High School Students

By high school, students have a fairly solid foundation in understanding human body systems such as the digestive, circulatory, and respiratory systems and recognize viruses as agents of infection. However, they may not have as clear an understanding of the immune system and thus have difficulty with understanding mechanisms and processes associated with infectious diseases.

Administering the Probe

This probe is appropriate at all grade levels. The last distracter on the list—"imbalance of body fluids"—comes from a predominant historical

belief that subsequently led people to treat illness by inducing vomiting, bleeding, or purging in order to adjust body fluids. As this phrase may be unfamiliar to younger students, consider eliminating it from the list when used with younger children. For older students who can distinguish between different types of microbes, you might consider deleting "germs" and adding two responses—"viruses" and "bacteria."

Related Ideas in *National Science Education Standards* (NRC 1996)

K–4 Personal Health

★ Individuals have some responsibility for their own health. Students should engage in personal habits—dental hygiene, cleanliness, and exercise—that will maintain and improve their health. At this level, children should come to understand how communicable diseases, such as colds, are transmitted and that some of the body's defense mechanisms prevent or overcome transmission.

5–8 Structure and Function in Living Systems

★ Disease is a breakdown in structures or functions of an organism. Some diseases are the result of intrinsic failures of the system. Others are the result of damage by infection by other organisms.

9–12 Personal and Community Health

• The severity of disease symptoms is dependent on many factors, such as human resistance and the virulence of disease-producing organisms. Many diseases can be prevented, controlled, or cured.

Related Ideas in *Benchmarks for Science Literacy* (AAAS 1993 and 2008)

Note: Benchmarks revised in 2008 are indicated by (R). New benchmarks added in 2008 are indicated by (N).

K–2 Physical Health

• Eating a variety of healthful foods and getting enough exercise and rest help people to stay healthy.

★ Some diseases are caused by germs, and some are not. Diseases caused by germs may be spread by people who have them. Washing one's hands with soap and water reduces the number of germs that can get into the body or that can be passed on to other people.

3–5 Physical Health

• Some germs may keep the body from working properly. For defense against germs, the human body has tears, saliva, and skin to prevent many germs from getting into the body and special cells to fight germs that do get into the body.

★ Indicates a strong match between the ideas elicited by the probe and a national standard's learning goal.

6–8 Physical Health

★ Viruses, bacteria, fungi, and parasites may infect the human body and interfere with normal body functions. A person can catch a cold many times because there are many varieties of cold viruses that cause similar symptoms.

• Specific kinds of germs cause specific diseases. (N)

6–8 Discovering Germs

★ Throughout history, people have created explanations for disease. Some have held that disease has spiritual causes, but the most persistent biological theory over the centuries was that illness resulted from an imbalance in the body fluids. The introduction of germ theory by Louis Pasteur and others in the 19th century led to the modern belief that many diseases are caused by microorganisms—bacteria, viruses, yeasts, and parasites.

Related Research

• The folklore about how an individual "catches" a common cold is very tenacious. The condition is not regarded as a disease, and the word *cold* reinforces the connection with environmental causes (Driver et al. 1994, p. 56).

• In a study by Brumby, Garrard, and Auman (1985), some students saw health and illness as two different concepts with different causes rather than as a continuum. Another sample of students saw illness as the negative end of a health continuum

of "lifestyle diseases" with no mention of infectious diseases (Driver et al. 1994).

• Exposure to TV and publicity on AIDS might influence modern children's ideas about infectious disease and predispose them more toward the germ theory of disease (Driver et al. 1994).

• Students have been known to hold conflicting ideas concurrently—at the same time, for example, believing that "all diseases are caused by germs" and that you can "catch a cold by getting cold and wet" (Driver et al. 1994).

• Research suggests that children often think of disease and decay as properties of the objects affected. They do not appear to hold a concept of microbes as agents of change (Driver et al. 1994, p. 55).

Suggestions for Instruction and Assessment

• When teaching about infectious diseases, distinguish among cause, transmission, and factors that lower resistance to disease.

• Engage older students in a debate regarding the many myths of the common cold. Encourage students to use their knowledge of cells, the immune system, and personal health to back up their claims with evidence.

• Use the story of Louis Pasteur and his contribution to the development of the germ theory. This historical episode is particularly relevant at the middle school level. In addition to tracing the development of ideas related to infectious diseases, it pro-

---

★ Indicates a strong match between the ideas elicited by the probe and a national standard's learning goal.

vides an excellent opportunity to highlight the nature of science.

Related NSTA Science Store Publications, NSTA Journal Articles, NSTA SciGuides, NSTA SciPacks, and NSTA Science Objects

American Association for the Advancement of Science (AAAS). 2001. *Atlas of science literacy.* Vol. 1. (See "Diseases," pp. 86–87.) Washington, DC: AAAS.

American Association for the Advancement of Science (AAAS). 2007. *Atlas of science literacy.* Vol. 2. (See "Discovering Germs," pp. 86–87.) Washington, DC: AAAS.

Pea, C., and D. Sterling. 2002. Cold facts about viruses. *Science Scope* (Nov./Dec.): 12–17.

Roy, K. 2003. Handwashing: A powerful preventative practice. *Science Scope* (Oct.): 12–14.

Sullivan, M. 2004. Career of the month: An interview with microbiologist Dale B. Emeagwali. *The Science Teacher* (Mar.): 76.

Related Curriculum Topic Study Guides

(Keeley 2005)

"Infectious Disease"

"Health and Disease"

"Personal and Community Health"

References

American Association for the Advancement of Science (AAAS). 1993. *Benchmarks for science literacy.* New York: Oxford University Press.

American Association for the Advancement of Science (AAAS). 2008. Benchmarks for science literacy online. *www.project2061.org/publications/bsl/online*

Brumby, M., J. Garrard, and J. Auman. 1985. Students' perceptions of the concept of health. *European Journal of Science Education* 7(3): 307–323.

Driver, R., A. Squires, P. Rushworth, and V. Wood-Robinson. 1994. *Making sense of secondary science: Research into children's ideas.* London: RoutledgeFalmer.

Keeley, P. 2005. *Science curriculum topic study: Bridging the gap between standards and practice.* Thousand Oaks, CA: Corwin Press.

National Research Council (NRC). 1996. *National science education standards.* Washington, DC: National Academy Press.

Digestive System

Six friends were talking about the function of the digestive system. This is what they said:

Mina: "I think the main function is to release energy from food."

Manny: "I think the main function is to help us breathe."

Sasha: "I think the main function is to break food down into molecules that can be absorbed by cells."

Harriet: "I think the main function is to break food down in the stomach into small pieces of food that can be used by the body."

Todd: "I think the main function is to carry bits of food and nutrients to all the different parts of our body."

Curtis: "I think the main function is to store food so that we can get energy when we need it."

Which student do you most agree with? Explain your thinking. Describe your ideas about the main function of the digestive system.

_____

_____

_____

_____

_____

_____

_____ .

_____

Digestive System

Teacher Notes

Purpose

The purpose of this assessment probe is to elicit students' ideas about the digestive system. The probe is designed to find out whether students realize a main function of the digestive system is to break food down into molecules that can be used by cells.

Related Concepts

cells, digestive system, food, nutrients

Explanation

The best answer is Sasha's: "I think the main function is to break food down into molecules that can be absorbed by cells." There is no single purpose of the digestive system; rather, it has two major purposes: (1) to break down food and (2) to prepare nutrients for absorption by cells. The digestive system carries out six basic functions: taking food in (ingestion), secretion, movement of food and wastes, breakdown of food, absorption from gastrointestinal tract to cells, and removal of wastes. In regard to Mina's response, the digestive system does not release energy from food. Instead it breaks food down into molecules that are absorbed by cells and that can then be used to release energy within the cell. Harriet's response is partially correct. The stomach does break ingested food down into smaller pieces of food. However, there is more to digestion than what happens in the stomach. These small pieces of food are not used directly by the body but are further broken down into small particles (molecules) absorbed by cells as they pass through the intestines. The difference between Sasha's and Harriet's responses is that the broken down food must be small enough to be taken in by

and used by cells. Even though the mouth and stomach break food down into small pieces of food, it is not until the food is broken down into the molecular units that make up the food that it can be used by the body to carry out the life processes that happen within cells. Todd is partially correct in that the digestive system does move food and nutrients; however, it moves these things through the digestive tract and not through different parts of the body. It is the circulatory system that moves nutrients throughout the body to cells.

Curricular and Instructional Considerations

Elementary Students

In the elementary grades, students learn basic ideas about the human body and body structures that help us take in and digest food, such as the mouth, teeth, and stomach. Young children primarily equate the stomach as the organ responsible for digesting food as they haven't yet learned how all the parts work together. By third grade, students begin to view the body as a system that works together and they can further explore what happens to food when it is taken into the body. At the upper-elementary level, it is important for students to know that food is broken down to obtain energy and materials for growth and repair, but the molecular aspect can wait until middle school.

Middle School Students

In the middle grades, students develop a more sophisticated understanding of the human

body and the organs and systems that work together to enable humans and other organisms to carry out their life processes. This is the time for them to understand that when food is broken down, it must be digested into molecules that can be absorbed and transported to different parts of the body. At this level, they are ready to understand the link between the digestive system and the circulatory system for breaking down food into molecules and transporting nutrients and to understand the role of the digestive and excretory system in eliminating the parts of food that are not used.

High School Students

Students at the high school level expand their understanding to encompass molecular energy release and the biochemical details related to metabolism. Their growing knowledge of cells helps them understand how molecules are taken into cells and used to carry out life processes.

Administering the Probe

If this probe is used with elementary students, consider substituting the word *molecules* with *tiny particles*.

Related Ideas in *National Science Education Standards* (NRC 1996)

K–4 Characteristics of Organisms

- Organisms have basic needs. For example, animals need air, water, and food; plants require air, water, nutrients, and light.

5–8 Structure and Function in Living Systems

- Cells carry on the many functions needed to sustain life. This requires that they take in nutrients, which they use to provide energy for the work that cells do and to make the materials that a cell or an organism needs.

- The human organism has systems for digestion, respiration, reproduction, circulation, excretion, movement, control and coordination, and protection from diseases. These systems interact with one another.

9–12 The Cell

- Most cell functions involve chemical reactions. Food molecules taken into cells react to provide the chemical constituents needed to synthesize other molecules. Both breakdown and synthesis are made possible by a large set of protein catalysts, called *enzymes*. The breakdown of some of the food molecules enables the cell to store energy in specific chemicals that are used to carry out the many functions of the cell.

Related Ideas in *Benchmarks for Science Literacy* (AAAS 1993 and 2008)

. . . .

Note: Benchmarks revised in 2008 are indicated by (R). New benchmarks added in 2008 are indicated by (N).

K–2 Basic Functions

- The human body has parts that help it seek, find, and take in food when it feels hunger—eyes and noses for detecting food, legs to get to it, arms to carry it away, and a mouth to eat it.

3–5 Basic Functions

- From food, people obtain fuel and materials for body repair and growth. The indigestible parts of food are eliminated. (R)

6–8 Basic Functions

- Organs and organ systems are composed of cells and help to provide all cells with basic needs.

- ★ For the body to use food for energy and building materials, the food must first be digested into molecules that are absorbed and transported to cells.

Related Research

- Some upper-elementary students have a primitive notion of the digestive system as the place where "lumps of food" are broken down, juices or acids dissolve the food, and "goodness" is somehow extracted from it. Few children ages 9–12 know that after food is changed in the stomach it is then broken down into even simpler substances that are carried to tissues throughout the body (Driver et al. 1994).

- Lower-elementary students know food is related to growing and being strong and healthy, but they are not aware of the physiological mechanisms (AAAS 1993).

- In a study conducted by Mintzes (1984), the youngest children up to age seven related the stomach to breathing, blood,

---

★ Indicates a strong match between the ideas elicited by the probe and a national standard's learning goal.

strength, and energy. When they are about seven years old, children begin to understand that the stomach helps to break down or digest food and that later the food is transferred somewhere else after being in the stomach (Driver et al. 1994).

Suggestions for Instruction and Assessment

- Students know that humans need food for energy and that the digestive system breaks down food so it can be used, but students often think that the food is just broken into smaller pieces. At the middle level, students should be explicitly taught the idea that food must be broken down into molecules in order for the body to use it.

- Show the connections among the digestive, circulatory, and respiratory systems in terms of breaking down food into molecules, transporting the molecules to cells, and releasing energy in the cells.

- Help students understand that the energy we get from food is released inside a cell. To get into the cell, the food must be broken down into molecules in order to pass through the cell membrane. Teaching students that food must eventually get into the cell in a form the cell can use will help them understand that food must be broken down into molecules.

- Use models and diagrams so that younger students can see that food continues to pass through the digestive system after the stomach. Some students think food is broken down in the stomach and the rest

passes out as waste. Few students comprehend the role of the intestines in terms of absorption.

Related NSTA Science Store Publications, NSTA Journal Articles, NSTA SciGuides, NSTA SciPacks, and NSTA Science Objects

Crowley, J. 2004. Nutritional chemistry. *The Science Teacher* (Apr.): 49–51.

Robertson, W. 2006. Science 101: How does the human body turn food into useful energy? *Science & Children* (Mar.): 60–61.

Schroeder, C. 2007. Inquiring into the digestive system. *Science Scope* (Nov.): 30–34.

Texley, J. 2001. Anatomy by logic. *Science Scope* (Sept.): 56–59.

Related Curriculum Topic Study Guides

(Keeley 2005)

"Human Body Systems"

"Food and Nutrition"

References

American Association for the Advancement of Science (AAAS). 1993. *Benchmarks for science literacy.* New York: Oxford University Press.

American Association for the Advancement of Science (AAAS). 2008. Benchmarks for science literacy online. *www.project2061.org/publications/bsl/online*

Driver, R., A. Squires, P. Rushworth, and V. Wood-Robinson. 1994. *Making sense of secondary science: Research into children's ideas.* London: RoutledgeFalmer.

Keeley, P. 2005. *Science curriculum topic study: Bridging the gap between standards and practice.* Thousand Oaks, CA: Corwin Press.

Mintzes, J. 1984. Naive theories in biology: Chil-dren's concepts of the human body. *School Science and Mathematics* 84 (7): 548–555.

National Research Council (NRC). 1996. *National science education standards.* Washington, DC: National Academy Press.

Camping Trip

Five friends were camping in the north woods. It was a clear night with mild weather conditions. Sunset was at 9:14 p.m. Sunrise was at 5:22 a.m. The five friends wondered when it would be coldest as they slept under the stars. This is what they said:

Colin: "I think it will be coldest right after the Sun sets."

Bono: "I think it will be coldest at midnight."

Jeri: "I think it will be coldest around 3:00 a.m."

Emma: "I think it will be coldest at the beginning of sunrise."

Kit: "I think it will be coldest a few hours after sunrise."

Which person do you agree most with and why? Explain your answer.

_____

_____

_____

_____

_____

_____

_____

_____

19

Camping Trip

Teacher Notes

Purpose

The purpose of this assessment probe is to elicit students' ideas about the effect of solar radiation on Earth's temperature. The probe is designed to find out whether students realize the Earth continues to cool after sundown and up to sunrise until there is sufficient radiation to begin warming the Earth.

Related Concepts

heat transfer, solar radiation, temperature, weather

Explanation

The best answer is Emma's: "I think it will be coldest at the beginning of sunrise." The coldest part of the day is generally right around the dawn, actually right after the sunrise while the Sun is still very low on the horizon. During

the night, the Earth's surface radiates the heat it has absorbed back out into space, allowing the temperature to drop. It does this during the day as well, but at night it has had the most time to radiate heat back to space without the incoming Sun's warmth to offset or compensate for the heat loss. Generally during the day more radiant energy is gained than lost and the Earth warms up. Between sunset and sunrise on a clear night, the Earth's surface generally receives no solar heat and steadily radiates heat back into space and thus cools. The temperature of the Earth's surface and the air in contact with it drops. Because the Sun is so low to the horizon at the beginning of sunrise, the solar radiation is very weak and is not yet strong enough to offset or compensate for all the heat escaping from the Earth. Clear skies prevent the heat rising from the Earth's sur-

face to be radiated back to Earth. As a result, the Earth's surface continues to lose heat for a short time following sunrise, and the air temperature continues to fall. Eventually, as the Sun rises, its rays hit the Earth's surface under a larger angle and become more concentrated. Eventually the concentration of the Sun's rays becomes large enough to compensate for the heat loss. The heat gain–loss balance is shifted, and the air finally begins to warm up.

Curricular and Instructional Considerations

Elementary Students

In the elementary grades, students learn about the heating and cooling of materials. They observe that warm materials gradually lose heat when they are no longer in contact with a heat source. They learn that the Sun heats the Earth and that during the night the Sun is not shining on the Earth. They can observe and analyze daily temperature fluctuations, including how the temperature in their area is usually lower in the evening and rises again in the daytime.

Middle School Students

In the middle grades, students learn about the various ways heat travels. They develop a more sophisticated understanding of solar radiation and how different materials absorb energy. They can devise models to observe how an object cools down when a source of light and heat is no longer in contact with an object or material and how different materials lose heat at different rates. They can collect and use temperature data to analyze heating and cooling patterns in their local area during a 24-hour photoperiod.

High School Students

Students at the high school level expand their experiences in observing daily heating and cooling of a local area to understanding the Earth's radiant energy balance. They then apply this understanding to climate. They should recognize greenhouse gases and explain how they trap energy and reduce the cooling of the Earth. They should also recognize the effects of clouds, oceans, snow and ice cover, and position of mountain ranges on heating and cooling.

Administering the Probe

This probe can be adapted to fit the geographic locale of the students it is used with. For example, use sunset and sunrise times for a local area during a specific date.

Related Ideas in *National Science Education Standards* (NRC 1996)

K–4 Objects in the Sky

- The Sun provides the light and heat necessary to maintain the temperature of the Earth.

5–8 Transfer of Energy

- The Sun is a major source of energy for changes on the Earth's surface. The Sun loses energy by emitting light. A tiny fraction of that light reaches the Earth, transferring energy from the Sun to the

Earth. The Sun's energy arrives as light with a range of wavelengths consisting of visible light, infrared radiation, and ultraviolet radiation.

5–8 Earth in the Solar System

• The Sun is the major source of energy for phenomena on the Earth's surface.

9–12 Energy in the Earth System

• Earth systems have internal and external sources of energy, both of which create heat. The Sun is the major external source of energy.

★ Global climate is determined by energy transfer from the Sun at and near the Earth's surface. This energy transfer is influenced by dynamic processes, such as cloud cover and the Earth's rotation, and static conditions, such as the position of mountain ranges and oceans.

Related Ideas in *Benchmarks for Science Literacy* (AAAS 1993 and 2008)

Note: Benchmarks revised in 2008 are indicated by (R). New benchmarks added in 2008 are indicated by (N).

K–2 Energy Transformation

• The Sun warms the land, air, and water.

3–5 The Earth

• The weather is always changing and can be described by measurable quantities such as temperature, wind direction and speed, and precipitation. (N)

6–8 The Earth

★ The temperature of a place on the Earth's surface tends to rise and fall in a somewhat predictable pattern every day and over the course of a year. (N)

6–8 Energy Transformation

• Heat can be transferred through materials by the collisions of atoms or across space by radiation.

9–12 The Earth

★ Weather (in the short run) and climate (in the long run) involve the transfer of energy in and out of the atmosphere. Solar radiation heats the land masses, oceans, and air.

9–12 Energy Transformation

• Whenever the amount of energy in one place diminishes, the amount of energy in other places or forms increases by the same amount. (R)

Related Research

• Our preliminary use of this probe revealed that many students from all grade levels think midnight is the coldest time of the night and that it slowly warms up after midnight.

• When we probed further with students who chose 3:00 a.m., they explained that it wouldn't be coldest at the beginning of sunrise because as soon as sunlight strikes the Earth, it starts warming up again.

★ Indicates a strong match between the ideas elicited by the probe and a national standard's learning goal.

Suggestions for Instruction and Assessment

- This probe lends itself to an inquiry investigation. Have students collect daily and nightly temperature data at hourly intervals during normal weather conditions, noting the time of sunrise and sundown to observe temperature patterns. Internet sites also provide hourly temperature data at specific locations on given dates as well as sunrise and sunset times. Use the P-E-O-E technique (Keeley 2008) to ask students to commit to a prediction that matches a response on the probe. Ask them to explain their thinking, make observations by collecting data, and then revise their explanations if their data do not support their predictions.

- Consider modeling this scenario using a material that absorbs heat from a heat lamp. Move the position and angle of the lamp to simulate the absorption of heat during the day and the slow loss of heat as the Sun sets over the horizon due to the rotation of the Earth. Turn off the lamp to simulate night. Simulate sunrise with a narrow angle between the material and the heat lamp, beginning with only part of the heat lamp shining on the material to simulate the Sun gradually rising over the horizon at dawn. Gradually widen the angle to simulate sunrise. Observe the cooling during the night and gradual rise in temperature right after sunrise.

Related NSTA Science Store Publications, NSTA Journal

Articles, NSTA SciGuides, NSTA SciPacks, and NSTA Science Objects

Childs, G. 2007. A solar energy cycle. *Science & Children* (Mar.): 26–29.

Damonte, K. 2005. Science shorts: Heating up, cooling down. *Science & Children.* (Jul.): 47–48.

Gilbert, S. W., and S. W. Ireton. 2003. *Understanding models in earth and space science.* Arlington, VA: NSTA Press.

Oates-Bockenstedt, C., and M. Oates. 2008. *Earth science success: 50 lesson plans for grades 6–9.* Arlington, VA: NSTA Press.

> **Related Curriculum Topic Study Guide**
> (Keeley 2005)
> "Weather and Climate"

References

American Association for the Advancement of Science (AAAS). 1993. *Benchmarks for science literacy.* New York: Oxford University Press.

American Association for the Advancement of Science (AAAS). 2008. Benchmarks for science literacy online. *www.project2061.org/publications/bsl/online*

Keeley, P. 2005. *Science curriculum topic study: Bridging the gap between standards and practice.* Thousand Oaks, CA: Corwin Press.

Keeley, P. 2008. *Science formative assessment: 75 practical strategies for linking assessment, instruction, and learning.* Thousand Oaks, CA: Corwin Press.

National Research Council (NRC). 1996. *National science education standards.* Washington, DC: National Academy Press.

Global Warming

Seven students argued about what they thought were major human causes of global warming. This is what they thought were causes that could be attributed to humans:

Maria: acid rain

Natalie: burning coal

Tessa: the fuel we use in our cars

Blaine: using leaded gasoline instead of unleaded

Anita: toxic chemicals in air pollution

Raul: the thinning of the Earth's ozone layer

Van: water pollution

Circle the name(s) of the student or students you agree with. Explain why you agree.

_____

_____

_____

_____

_____

_____

_____

_____

_____

_____

Global Warming

Teacher Notes

Purpose

The purpose of this assessment probe is to elicit students' ideas about global warming. The probe is designed to find out what students think contributes to global warming.

Related Concepts

climate change, global warming, greenhouse gas

Explanation

The best answers are from Natalie and Tessa. The major cause of human-induced global warming comes from our use of fossil fuels that produce greenhouse gases. As solar radiation reaches Earth, clouds and dust reflect some of this back toward space, while some solar radiation warms the surface of the Earth. As a result, the Earth emits infrared radiation into the atmosphere, where some of the

atmospheric gases, including water vapor, carbon dioxide, methane, and nitrous oxide, are capable of absorbing the longer wavelength radiant energy. These gases are collectively called greenhouse gases, which warm the atmosphere, resulting in a climate that keeps our planet warm and livable.

This greenhouse effect is natural and necessary for life as we know it. In recent years, however, there has been an increase in the anthropogenic (produced by humans) production of greenhouse gases. Global atmospheric concentrations of carbon dioxide, methane, and nitrous oxide have increased markedly as a result of human activities since 1750 and now far exceed preindustrial values (IPCC 2007). Carbon dioxide is the most significant anthropogenic greenhouse gas. The primary source of the increased atmospheric concentration of carbon

dioxide is a result of the use of fossil fuels, such as coal and gasoline, because burning of fossil fuels (hydrocarbons found in the top layer of the Earth's crust) produces carbon dioxide gas.

The increased concentration of greenhouse gases has resulted in an increased absorption of radiant energy in the atmosphere, resulting in a warming of the global climate system as evident from observations of increases in global average air and ocean temperatures, widespread melting of snow and ice, and rising global mean sea level (IPCC 2007).

There are a number of human-caused phenomena that are of significant environmental concern but are not major causes of global warming. Acid rain, caused by emissions of sulfur and nitrogen compounds that react in the atmosphere to produce acids, can affect a region's atmosphere, hydrosphere, geosphere, and biosphere. Air pollution and toxic chemicals can have similar impacts. Ozone depletion in the stratosphere may have some bearing on the chemistry of the atmosphere, but more important, it is linked to biological concerns—for example, skin cancer and cataracts, damage to plants, and reduction of plankton populations in the ocean—due to allowing more ultraviolet radiation to penetrate our atmosphere.

Curricular and Instructional Considerations

Elementary Students

In the elementary grades, students become familiar with energy sources and have oppor-tunities to learn about the Sun's energy and how it heats the Earth. Students keep daily records of temperature, looking for patterns, before they develop an understanding of climate. Because the issue of global warming is common in the media, students at this age begin developing early conceptions of what it is and what causes it.

Middle School Students

In the middle grades, students may confuse the concepts of energy and energy sources and benefit from experiences that focus on energy transformation with design challenges and energy conversion systems. They trace where energy comes from and goes, using examples that involve different forms of energy such as heat and light. They begin to understand the laws of nature and to become aware that although energy can be transformed, it is not destroyed. Students connect their prior knowledge of the Earth and its weather to an understanding of climate. Their geometric reasoning and experiences with scale help them to shift their frame of reference away from the Earth's surface and to examine the effects of radiation at different angles to explain the seasons. They examine issues related to the impact of technology, such as burning fossil fuels. They analyze the causes and formulate ideas for solutions.

High School Students

Students at the high school level compare industrial and nonindustrial societies by their standards of living and energy consumption. They examine the consequences of the

world's dependence on fossil fuels, explore a wide range of alternative energy resources and technologies, consider trade-offs in each, and propose policies for conserving and managing limited energy resources. Principles of radiation, energy, and conservation are integrated into their views of Earth systems science. Their growing knowledge of chemistry helps them understand what greenhouse gases are and their impact on the atmosphere. A consciousness is developed that important decisions we make about our daily lives sometimes require us to make compromises; societal impacts are studied as well as those that affect human health and the environment.

Administering the Probe

This probe is best used at the middle and high school levels, particularly if students have been previously exposed to the terms *climate change* and *global warming*. You may want to encourage students to explain not only why they agree with the choice(s) they selected but also why they did not select the other choices.

Related Ideas in *National Science Education Standards* (NRC 1996)

. . . .

K–4 Changes in Environments

- Changes in environments can be natural or influenced by humans. Some changes are good, some are bad, and some are neither good nor bad.

5–8 Structure of the Earth System

- The atmosphere is a mixture of nitrogen, oxygen, and trace gases that include water vapor. The atmosphere has different properties at different elevations.

5–8 Earth in the Solar System

- The Sun is the major source of energy for phenomena on the Earth's surface.

5–8 Understandings About Science and Technology

- Technological solutions have intended benefits and unintended consequences. Some consequences can be predicted, others cannot.

9–12 Energy in the Earth System

- Earth systems have internal and external sources of energy, both of which create heat. The Sun is the major external source of energy.
- Global climate is determined by energy transfer from the Sun at and near the Earth's surface. This energy transfer is influenced by dynamic processes, such as cloud cover and the Earth's rotation, and static conditions. such as the position of mountain ranges and oceans.

9–12 Science in Personal and Social Perspectives

- Materials from human societies affect both physical and chemical cycles of the Earth.

Related Ideas in *Benchmarks for Science Literacy* (AAAS 1993 and 2008)

Note: Benchmarks revised in 2008 are indicated by (R). New benchmarks added in 2008 are indicated by (N).

K–2 Energy Sources and Use

• People burn fuels such as wood, oil, coal, or natural gas, or use electricity to cook their food and warm their houses.

K–2 Energy Transformations

• The Sun warms the land, air, and water.

3–5 The Earth

• Air is a substance that surrounds us and takes up space. It is also a substance whose movement we feel as wind.

6–8 Energy Sources and Use

• Different ways of obtaining, transforming, and distributing energy have different environmental consequences.

★ By burning fuels, people are releasing large amounts of carbon dioxide into the atmosphere and transforming chemical energy into thermal energy, which spreads throughout the environment.

6–8 Processes That Shape the Earth

• Human activities such as reducing the amount of forest cover, increasing the amount and variety of chemicals released into the atmosphere, and intensive farm- ing have changed the Earth's land, oceans, and atmosphere.

9–12 Energy Sources and Use

• At present, all fuels have advantages and disadvantages so society must consider the trade-offs among them.

• Industrialization brings an increased demand for and use of energy. Such usage contributes to the high standard of living in the industrially developing nations but also leads to more rapid depletion of the Earth's energy resources and to environ- mental risks associated with the use of fos- sil and nuclear fuels.

9–12 The Earth

• Weather (in the short run) and climate (in the long run) involve the transfer of energy in and out of the atmosphere. Solar radia- tion heats the land masses, oceans, and air. Transfer of heat energy at the boundaries between the atmosphere, the land masses, and the oceans results in layers of differ- ent temperatures and densities in both the ocean and atmosphere.

★ Greenhouse gases in the atmosphere, such as carbon dioxide and water vapor, are transparent to much of the incoming sun- light but not to the infrared light from the warmed surface of the Earth. When green- house gases increase, more thermal energy is trapped in the atmosphere, and the temper- ature of the Earth increases the light energy radiated into space until it again equals the light energy absorbed from the Sun. (N)

★ Indicates a strong match between the ideas elicited by the probe and a national standard's learning goal.

★ The Earth's climates have changed in the past, are currently changing, and are expected to change in the future, primarily due to changes in the amount of light reaching places on the Earth and the composition of the atmosphere. The burning of fossil fuels in the last century has increased the amount of greenhouse gases in the atmosphere, which has contributed to Earth's warming.

Related Research

- Students of all ages may confuse the ozone layer with the greenhouse effect and may have a tendency to imagine that all environmentally friendly actions help to solve all environmental problems (e.g., that the use of unleaded gas reduces the risk of global warming [AAAS 2007]).

- Students have difficulty linking relevant elements of knowledge when explaining the greenhouse effect and may confuse the natural greenhouse effect with the enhancements of that effect (AAAS 2007).

- Students may identify the transfer of energy to the environment with pollution or waste materials being thrown into the environment (AAAS 1993).

- Some students confuse lead and other types of pollution with the greenhouse effect (Driver et al. 1994).

Suggestions for Instruction and Assessment

- Discuss the distinction between the natural greenhouse effect and global warming,

highlighting the ability of greenhouse gases to trap infrared radiation as natural and necessary for life as we know it. It is the increase in concentration of greenhouse gases that is producing the global warming effect, resulting in an increase in global average air and ocean temperatures.

- Hold a climate-change town hall meeting. Have students assume a variety of citizen and government agency roles and discuss the factors contributing to global warming. Encourage the group to make decisions that would reduce the anthropogenic production of greenhouse gases, considering the personal, national, and global impacts and the economic, societal, and scientific costs and benefits. Their goal should be to reach a consensus on how to reduce the consumption in a realistic way, not to debate and have a "loser" and a "winner."

- Encourage students to consider ways they can "think globally, act locally." Identify steps they can take to reduce their carbon footprint. Use a carbon dioxide emissions calculator. These and other global climate change education resources can be found at the American Association for the Advancement of Science website: *www.aaas.org/climate*.

- Include understandings of the nature of science as you explore complex issues of global warming and climate change. The Intergovernmental Panel on Climate Change (IPCC) was established to bring an international group of scientists together

---

★ Indicates a strong match between the ideas elicited by the probe and a national standard's learning goal.

in order to evaluate the risk of climate change caused by human activity. For more information, go to *www.ipcc.ch*. Examine, analyze, and interpret the data that have been summarized in the *Climate Change 2007: Synthesis Report* (IPCC 2007).

Related NSTA Science Store Publications, NSTA Journal Articles, NSTA SciGuides, NSTA SciPacks, and NSTA Science Objects

Constible, J., L. Sandro, and R. E. Lee, Jr. 2007. A cooperative classroom investigation of climate change. *The Science Teacher* (Sept.): 56–63.

Environmental Literacy Council and National Science Teachers Association (NSTA). 2007. *Global climate change: Resources for environmental literacy.* Arlington, VA: NSTA Press.

Environmental Literacy Council and National Science Teachers Association (NSTA). 2007. *Resources for environmental literacy: Five teaching modules for middle and high school teachers.* Arlington, VA: NSTA Press.

Miller, R. G. 2006. Issues in depth: Inside global warming. *Science Scope* (Oct.): 56–60.

National Science Teachers Association (NSTA). 2007. The ocean's effect on weather and climate. NSTA SciPack. Online at *http://learningcenter.nsta.org/product_detail.aspx?id=10.2505/6/SCP-OCW.*

Smith, Z. 2007. A record of climate change. *The Science Teacher* (Sept.): 48–53.

Space, W. 2007. Climate physics: Using basic physics concepts to teach about climate change. *The Science Teacher* (Sept.): 44–47.

Related Curriculum Topic Study Guide

(Keeley 2005)
"Weather and Climate"

References

American Association for the Advancement of Science (AAAS). 1993. *Benchmarks for science literacy.* New York: Oxford University Press.

American Association for the Advancement of Science (AAAS). 2007. *Atlas of science literacy.* Vol. 2. (See "Weather and Climate," pp. 20–21.) Washington, DC: AAAS.

American Association for the Advancement of Science (AAAS). 2008. Benchmarks for science literacy online. *www.project2061.org/publications/bsl/online*

Driver, R., A. Squires, P. Rushworth, and V. Wood-Robinson. 1994. *Making sense of secondary science: Research into children's ideas.* London: RoutledgeFalmer.

Intergovernmental Panel on Climate Change (IPCC). 2007. *Climate change 2007: Synthesis report.* Geneva, Switzerland: IPCC.

Keeley, P. 2005. *Science curriculum topic study: Bridging the gap between standards and practice.* Thousand Oaks, CA: Corwin Press.

National Research Council (NRC). 1996. *National science education standards.* Washington, DC: National Academy Press.

Where Does Oil Come From?

Oil is an important energy resource used by humans. Several friends were arguing about where this energy resource came from. This is what they said:

Julie: "It came mostly from fossil remains of giant ferns and trees that lived millions of years ago."

Ross: "It came mostly from inside ancient rocks that melted inside the Earth millions of years ago."

Delores: "It came mostly from shallow ocean water that changed into oil after millions of years."

Edie: "It came mostly from a gooey liquid that was inside ancient volcanoes millions of years ago."

Nathan: "It came mostly from the remains of dinosaurs that decayed millions of years ago."

Seth: "It came mostly from microscopic and other ocean organisms millions of years ago."

Justine: "It came mostly from ancient mud, sand, and soil that eventually turned to liquid inside the Earth millions of years ago."

Malia: "It came mostly from gasoline that was trapped inside the Earth's crust for millions of years."

Cecelia: "It came mostly from the rotting blubber of ancient whales that lived millions of years ago."

Circle the name of the person you most agree with. Explain your thinking. Describe where you think oil came from and how it was formed.

Where Does Oil Come From?

Teacher Notes

Purpose

The purpose of this assessment probe is to elicit students' ideas about an important fossil fuel used by humans. The probe is designed to reveal how students trace oil back to its original source of material.

Related Concepts

fossil fuel, natural resource, nonrenewable resource

Explanation

The best answer is Seth's: "It came mostly from microscopic and other ocean organisms millions of years ago." The majority of petroleum oil is generally thought to come from the fossil remains of tiny animal and plant-like marine organisms. The majority of these organisms were phytoplankton and zooplank-

ton. Larger plants and animals may have also contributed to the formation of oil, but their contribution is fairly insignificant compared with the tiny marine organisms; therefore, Nathan's response would not be correct. As these tiny sea organisms died, their bodies collected on the seafloor and were gradually buried under layers of sediment and subsequent rock layers. Justine's response mentions the sediments but not the organisms contained in those sediments. As the accumulating sediments exerted pressure and heat, the remains of the organisms were gradually chemically transformed over millions of years into oil. Julie's response applies to coal, another type of fossil fuel. Coal is formed through a similar process involving accumulation of dead organisms and conditions of high heat and pressure, but coal comes primarily from land

vegetation—trees and giant ferns that lived long ago, died, and accumulated faster than they could decompose. The word *petroleum* means "rock oil," which may imply to some people that oil comes primarily from rocks rather than once living organisms, hence Ross's response. Malia's response refers to a product of oil—gasoline. Oil is refined and processed to make gasoline. Gasoline is not found naturally within the Earth like oil.

Curricular and Instructional Considerations

Elementary Students

In the elementary grades, students learn about different Earth materials that provide resources for human use. They learn that some materials are renewable and others cannot be renewed and are limited. As they learn about fossils, they can connect the idea of fossils to fossil fuels, simply understanding that some fuels, like oil, coal, and natural gas, come from the remains of ancient plants and animals.

Middle School Students

In the middle grades, students develop a more sophisticated understanding of fossil fuels by learning about the geologic processes that formed them and the differences among fossil fuels. Their increased comprehension of prehistoric conditions and the magnitudes of geologic time helps them understand why fossil fuels are considered to be nonrenewable resources.

High School Students

Students at the high school level build upon their K–8 knowledge of fossil fuels by recognizing why these fuels are such rich sources of energy. Their knowledge of chemistry helps them understand how the energy released from breaking the chemical bonds of oil and coal's hydrocarbon molecules came originally from the solar energy captured by phytoplankton and plants and is released through the combustion process.

Administering the Probe

Make sure students know that the probe is referring to petroleum oil, not other forms of oil they may be thinking of, such as vegetable oil, olive oil, or mineral oil. For younger students, consider reducing the choices to four or five answers to select from.

Related Ideas in *National Science Education Standards* (NRC 1996)

K–4 Properties of Earth Materials

* Earth materials provide many of the resources that humans use.

K–4 Types of Resources

* Resources are things we get from the living and nonliving environment to meet the needs and wants of a population.
* Some resources are basic materials, such as air, water, and soil; some are produced from basic resources such as food, fuel, and building materials.

- The supply of many resources is limited.

5–8 Structure of the Earth System
- Living organisms have played many roles in the Earth system, including producing some types of rocks.

5–8 Earth's History
- Fossils provide important evidence of how life and environmental conditions have changed.

9–12 Geochemical Cycles
- The Earth is a system containing essentially a fixed amount of each stable chemical atom or element. Each element can exist in several different chemical reservoirs. Each element on Earth moves among reservoirs in the solid Earth, oceans, atmosphere, and organisms as part of geochemical cycles.

9–12 Natural Resources
- Human populations use resources in the environment in order to maintain and improve their existence.
- The Earth does not have infinite resources. Increasing human consumption places severe stress on the natural processes that renew some resources, and it depletes those resources that cannot be renewed.

Related Ideas in *Benchmarks for Science Literacy* (AAAS 1993 and 2008)

. . . .

Note: Benchmarks revised in 2008 are indicated by

(R). New benchmarks added in 2008 are indicated by (N).

K–2 Energy Sources and Use
- People burn fuels such as wood, oil, coal, or natural gas or use electricity to cook their food and warm their houses.

3–5 Flow of Matter and Energy
- Over the whole Earth, organisms are growing, dying, and decaying, and new organisms are being produced by the old ones.

6–8 Energy Sources and Use
- Some resources are not renewable or renew very slowly. Fuels already accumulated in the Earth, for instance, will become more difficult to obtain as the most readily available resources run out. How long the resources will last, however, is difficult to predict. The ultimate limit may be the prohibitive cost of obtaining them. (N)

9–12 Flow of Matter and Energy
- ★ At times, environmental conditions are such that land and marine organisms reproduce and grow faster than they die and decompose to simple carbon-containing molecules that are returned to the environment. Over time, layers of energy-rich organic material inside the Earth have been chemically changed into great coal beds and oil pools. (R)

9–12 Energy Sources and Use
- ★ Sunlight is the ultimate source of most of the energy we use. The energy in fos-

---

★ Indicates a strong match between the ideas elicited by the probe and a national standard's learning goal.

sil fuels such as oil and coal comes from energy that plants captured from the Sun long ago. (N)

9–12 The Earth

- The Earth has many natural resources of great importance to human life. Some are readily renewable, some are renewable only at great cost, and some are not renewable at all. (N)

Related Research

- Some middle school students think dead organisms simply rot away. They do not realize that the matter from the dead organism is converted into other materials in the environment (AAAS 1993, p. 343). Also, some middle school students think organisms and materials in the environment are very different types of matter (AAAS 1993, p. 342). Because of these notions, some students may have difficulty believing that oil and coal were once the remains of living organisms.

- Misconceptions about fossil fuels may arise for a variety of reasons. Students sometimes misinterpret the word *fossil* to mean dinosaur, and sometimes animals like dinosaurs are used in cartoons that depict the formation of fossil fuels (also note the logo used on Sinclair gas stations). Some students think fossil fuel comes from whale blubber because they know that, historically, whale oil was used to heat and light homes before petroleum oil was discovered in the ground (Rule 2005).

- In a study comparing gifted with average students, both sets of students had similar misconceptions, with little difference between numbers of students who held misconceptions in both groups (Rule 2005).

Suggestions for Instruction and Assessment

- Help students understand what environmental conditions and the types of life-forms that existed hundreds of millions of years ago were like at that time. Some students think the Earth today is the same as it has always been. Learning about the changes that happen over long periods of time, including changes that happen on the seafloor, will help them understand the long geologic process that contributes to the formation of oil.

- When teaching about how one type of rock can be transformed metamorphically into a different type of rock under conditions of intense heat and pressure over long periods of time, include information about how fossil fuels were also formed by heat and pressure over long periods of time, transforming material from living organisms into coal or oil.

- Challenge students with the question, "If time, heat, and pressure are what it takes to create oil from dead microscopic plants and animals, why can't we just manufacture oil to replenish the supply inside the Earth?"

- A Java simulation showing how oil is formed can be found at *www.sciencelearn. org.nz/contexts/future_fuels/sci_media/ animations/oil_formation*.

- The persistence of fossil fuel misconceptions into adulthood indicates the critical need to help younger students understand the scientific concepts involved. Students need to understand how fossil fuels originated in order to understand their nonrenewable nature, their uneven distribution worldwide, and the ensuing political consequences (Rule 2005).

Related NSTA Science Store Publications, NSTA Journal Articles, NSTA SciGuides, NSTA SciPacks, and NSTA Science Objects

Energy Resources. NSTA SciGuide. Online at *http://learningcenter.nsta.org/product_detail. aspx?id=10.2505/5/SG-08*

Hudson, T., and G. Camphire. 2005. Petroleum and the environment. *The Science Teacher* (Dec.): 34–35.

Related Curriculum Topic Study Guides

(Keeley 2005)

"Energy Resources and Use"

"Earth's Natural Resources"

References

American Association for the Advancement of Science (AAAS). 1993. *Benchmarks for science literacy.* New York: Oxford University Press.

American Association for the Advancement of Science (AAAS). 2008. Benchmarks for science literacy online. *www.project2061.org/publications/ bsl/online*

Keeley, P. 2005. *Science curriculum topic study: Bridging the gap between standards and practice.* Thousand Oaks, CA: Corwin Press.

National Research Council (NRC). 1996. *National science education standards.* Washington, DC: National Academies Press.

Rule, A. C. 2005. Elementary students' ideas concerning fossil fuel energy. *Journal of Geoscience Education* 53 (3): 309–18.

Where Would It Fall?

Six friends were talking about asteroids and meteorites that could fall to the Earth. The friends wondered where an object from space would most likely fall. This is what they said:

Maya: "I think it has the greatest chance of landing in a desert."

Elsa: "I think it is most apt to land where humans are living."

Walter: "I think it will most likely land in an ocean."

Mac: "I think it will probably land on an ice-covered area."

Amber: "Chances are it will land on the largest continent."

Evan: "Most likely it will land in a body of freshwater."

Which person do you most agree with? Explain your reasons for where you think a large object from space would most likely fall.

_____

_____

_____

_____

_____

_____

_____

_____

Where Would It Fall?

Teacher Notes

Purpose

The purpose of this assessment probe is to elicit students' ideas about the distribution of land, oceans, freshwater, and ice. The probe is designed to find out whether students realize that most of the Earth is covered by oceans.

Related Concepts

Earth history, Earth's water distribution, oceans, surface of the Earth

Explanation

The best answer is Walter's: "I think it will most likely land in an ocean." Water covers about 71% of the Earth's surface. Of this 71%, 97% of the water is in the oceans. The probability of the object landing in the ocean is about 69%. Freshwater from glaciers and ice caps covers about 2% of Earth's total water,

with the rest coming from groundwater, rivers, lakes, and other fresh surface waters. About 29% of the Earth's surface is covered by land. The largest continent is Asia, which makes up about 30% of the Earth's total land surface. Deserts cover about 20% of the Earth's land area. About 12% of Earth's total land area is inhabited by humans.

Curricular and Instructional Considerations

Elementary Students

At the elementary level, students develop an appreciation of Earth's limited resources. With a developing geographic perspective, they begin to see how much of the Earth is covered by water, although the scale of coverage is not well recognized until middle school.

Middle School Students

At the middle school level, students develop an increased understanding of the distribution of Earth's land and water resources. Their increased sense of geography and views of the Earth from space help them recognize the oceans as the predominant feature on Earth's surface.

High School Students

Students at this level have a greater understanding of the global importance of oceans and their distribution worldwide. As they investigate ocean currents and the oceans' effects on climate, they frequently encounter visualizations that help reinforce their knowledge of the distribution of land and water. However, because students live in a terrestrial environment, they may intuitively think most of the Earth is covered by land.

Administering the Probe

Make sure students understand the context— an object the size of a large rock, falling from the sky and landing somewhere on Earth. Don't show them a map of the Earth. The purpose of the probe is to examine whether they conceptually understand that most of the Earth is covered by ocean. Showing them a map of the world would clue them to the answer before you have a chance to probe for their ideas about land, ice, and water.

Related Ideas in *National Science Education Standards* (NRC 1996)

K–4 Properties of Earth Materials

• Earth materials are solid rocks and soils, water, and the gases of the atmosphere.

5–8 Structure of the Earth System

★ Water, which covers the majority of the Earth's surface, circulates through the crust, oceans, and atmosphere in what is known as the water cycle.

Related Ideas in *Benchmarks for Science Literacy* (AAAS 1993 and 2008)

Note: Benchmarks revised in 2008 are indicated by (R). New benchmarks added in 2008 are indicated by (N).

6–8 The Earth

★ The Earth is mostly rock. Three-fourths of the Earth's surface is covered by a relatively thin layer of water (some of it frozen), and the entire planet is surrounded by a relatively thin layer of air. (R)

Related Research

• Although we could find no formal research in this area, preliminary field-test results indicated that elementary students believe most of the Earth is covered by land inhabited by people. The notion that the oceans cover most of Earth's surface begins to

---

★ Indicates a strong match between the ideas elicited by the probe and a national standard's learning goal.

be more common in middle school and increases through high school.

Suggestions for Instruction and Assessment

- Have students examine a globe or map of the world to see that oceans cover the majority of our planet.
- An inflatable Earth "beach ball" can be used as a model to test this problem. Throw the beach ball 25 times or more. Have each student catch the ball with two hands, looking at what part of the globe the tip of their right index finger is touching. The tip of their index finger will indicate where the object lands on Earth. Have them call out "land" or "water" (include "ice" or "desert" if the ball shows Earth's features) each time the ball is caught, depending on what their finger is touching. Tally the students' responses. Probability indicates that most catches will land on water, primarily ocean.

Related NSTA Science Store Publications, NSTA Journal Articles, NSTA SciGuides, NSTA SciPacks, and NSTA Science Objects

Kelly, C. 2002. The diminishing apple. *Science & Children* (Feb.): 26–30.

Ford, B., and P. Smith. 2000. *Project Earth Science: Physical oceanography*. Arlington, VA: NSTA Press.

> ### Related Curriculum Topic Study Guides
>
> (Keeley 2005)
> "Water in the Earth System"
> "Structure of the Solid Earth"

References

American Association for the Advancement of Science (AAAS). 1993. *Benchmarks for science literacy.* New York: Oxford University Press.

American Association for the Advancement of Science (AAAS). 2008. Benchmarks for science literacy online. *www.project2061.org/publications/bsl/online*

Keeley, P. 2005. *Science curriculum topic study: Bridging the gap between standards and practice.* Thousand Oaks, CA: Corwin.

National Research Council (NRC). 1996. *National science education standards.* Washington, DC: National Academy Press.

Moonlight

Five friends noticed they could see better at night when there was a full Moon. They wondered where the moonlight came from. This is what they said:

Curtis: "The Moon reflects the light from the Earth."

Chet: "The light from the Sun bounces off the Moon."

Clarence: "The Moon gets its light from distant stars."

Fallon: "The Moon absorbs light from the Sun during the day."

Deirdre: "There is light inside of the Moon that makes it shine."

Which person do you most agree with? Explain your thinking about moonlight.

_____

_____

_____

_____

_____

_____

_____

_____

Moonlight

Teacher Notes

Purpose

The purpose of this assessment probe is to elicit students' ideas about light and the Moon. The probe is designed to find out what students think is the source of a full Moon's light.

Related Concepts

light reflection, Moon, Moon phases

Explanation

The best answer is Chet's: "The light from the Sun bounces off the Moon." In other words, the Moon reflects sunlight. Even though you do not see the Sun in the evening, it does shine on the surface of the Moon. The light from the Sun reflects off the Moon and travels to Earth. Therefore, during a full Moon, the moonlight that helps us see better during the evening after the Sun has set is actually sunlight that is bounc-

ing off the Moon, striking the Earth, reflecting off objects, and entering our eyes, which allows us to see things at night. When there is no full Moon, there is less light reaching and reflecting off the Earth; therefore, we see better when there is a full Moon. In contrast, stars emit their own light. They are sources of light rather than objects that reflect light.

Curricular and Instructional Considerations

Elementary Students

In the elementary grades, students observe different phases of the Moon over time and describe the monthly pattern. They should also be challenged to think about why we can see the Moon, developing the idea that light from the Sun is reflected by the Moon toward Earth.

Reflection of light is a prerequisite understanding that should be developed before students learn that the Moon reflects light.

Middle School Students

In the middle grades, students develop a more sophisticated understanding of light reflection and the phases of the Moon. Now is the time when they can begin putting together ideas about the Earth, Moon, and Sun system to construct an understanding of what causes the phases of the Moon. By middle school, they should be able to distinguish between stars that give off their own light versus Moons and planets that can be seen by the Sun's reflected light.

High School Students

By high school, students should understand that the light from the Moon comes from reflected sunlight. However, there may be some students who still hold on to early misconceptions that were never challenged.

Administering the Probe

Make sure students understand the context of this probe. Students in urban areas may have never experienced the difference in nighttime darkness during the light of a full Moon.

Related Ideas in *National Science Education Standards* (NRC 1996)

· ·

K–4 Objects in the Sky

★ The Sun, Moon, stars, clouds, birds, and airplanes all have properties, locations,

and movements that can be observed and described.

K–4 Light, Heat, Electricity, and Magnetism

· Light travels in a straight line until it strikes an object. Light can be reflected by a mirror, refracted by a lens, or absorbed by the object.

5–8 Transfer of Energy

★ Light interacts with matter by transmission (including refraction), absorption, or scattering (including reflection).

5–8 Earth in the Solar System

· Most objects in the solar system are in regular and predictable motion. Those motions explain such phenomena as the phases of the Moon.

Related Ideas in *Benchmarks for Science Literacy* (AAAS 1993 and 2008)

· ·

Note: Benchmarks revised in 2008 are indicated by (R). New benchmarks added in 2008 are indicated by (N).

K–2 The Universe

· The Moon looks a little different every day but looks the same again about every four weeks.

★ Indicates a strong match between the ideas elicited by the probe and a national standard's learning goal.

3–5 Motion

- Light travels and tends to maintain its direction of motion until it interacts with an object or material. Light can be absorbed, redirected, bounced back, or allowed to pass through. (N)

6–8 The Earth

★ The Moon's orbit around the Earth once in about 28 days changes what part of the Moon is lighted by the Sun and how much of that part can be seen from the Earth—the phases of the Moon.

Related Research

- Understanding the phases of the Moon is very challenging for students. To understand the phases of the Moon they must first master the idea of a spherical Earth and understand the concept of light reflection and how the Moon gets its light from the Sun (AAAS 1993).

Suggestions for Instruction and Assessment

- Compare and contrast objects that emit their own light and objects that reflect light, and connect this to how we see the Moon.
- For students who believe the Moon creates its own light, use a model to confront them with their ideas. Shine a light on a white Styrofoam ball in a dark room. Have students describe what they see. Ask them to examine the ball to decide if there is something in it that caused it to light up. Give students time to revise their explanation after experiencing the model.
- Model the relationship between the Moon and the Sun at night. Using a model, show how the Sun shines on the Moon, even though we do not see the Sun during the night.

Related NSTA Science Store Publications, NSTA Journal Articles, NSTA SciGuides, NSTA SciPacks, and NSTA Science Objects

Ansberry, K., and E. Morgan. 2008. Teaching through trade books: Moon phases and models. *Science & Children* (Sept.): 20–22.

Gilbert, S., and S. Ireton. 2003. *Understanding models in Earth and space science.* Arlington, VA: NSTA Press.

Young, T., and M. Guy. 2008. The Moon's phases and the self shadow. *Science & Children* (Sept.): 30–35.

Related Curriculum Topic Study Guides

(Keeley 2005)
"Earth, Moon, Sun System"
"Visible Light, Color, and Vision"

References

American Association for the Advancement of Science (AAAS). 1993. *Benchmarks for science literacy.* New York: Oxford University Press.

American Association for the Advancement of Science (AAAS). 2008. Benchmarks for science lit-

---

★ Indicates a strong match between the ideas elicited by the probe and a national standard's learning goal.

eracy online. *www.project2061.org/publications/ bsl/online*

Keeley, P. 2005. *Science curriculum topic study: Bridging the gap between standards and practice.* Thousand Oaks, CA: Corwin Press.

National Research Council (NRC). 1996. *National science education standards.* Washington, DC: National Academy Press.

Lunar Eclipse

People have been fascinated by lunar (Moon) eclipses for ages. For a time, the full Moon seems to disappear as it changes color, darkens, and then reappears. Throughout time, people have had different ideas about what causes a lunar eclipse. Here are some of their ideas:

A A nearby planet passes between the Earth and the Moon.

B The Sun passes between the Earth and the Moon.

C The Moon passes between the Sun and the Earth.

D The Earth passes between the Sun and the Moon.

E The clouds block out the Moon.

F A nearby planet's shadow falls on the Moon.

G The Moon's shadow falls on the Earth.

H The Moon turns to the dark side and then back to the light side.

Circle the idea you think best explains what causes a lunar (Moon) eclipse. Explain your thinking about lunar eclipses.

_____

_____

_____

_____

_____

_____

_____

Lunar Eclipse

Teacher Notes

Purpose

The purpose of this assessment probe is to elicit students' ideas about eclipses. The probe is designed to find out what students think causes a lunar eclipse.

Related Concept

lunar eclipse

Explanation

The best answer is D: The Earth passes between the Sun and the Moon. A lunar eclipse occurs whenever the Moon passes through some portion of the Earth's shadow. This occurs when the Sun, Earth, and Moon are in very close alignment, with the Earth in the middle. There is always a full Moon the night of a lunar eclipse. You will usually see a bright lunar disk gradually turn dark as it passes through Earth's shadow—sometimes a coppery red color—for as long as an hour or more during a total eclipse. The Earth's shadow has two parts, called the *penumbra* and the *umbra*. The umbra is much darker than the penumbra. When the entire Moon passes through the umbra, it is called a total lunar eclipse. When only part of the Moon passes through the umbra, it is called a partial lunar eclipse. And when the Moon goes through the penumbra, it is called a penumbral eclipse. In each of these types of eclipses, the Earth is always between the Sun and the Moon. Lunar eclipses (as well as solar eclipses) happen roughly every six months. A lunar eclipse is visible to anyone on Earth who would have otherwise seen the full Moon.

Curricular and Instructional Considerations

Elementary Students

In the elementary grades, students learn basic ideas about the Moon, including the fact that the Moon can be seen in the daytime and that it changes each day in a pattern that lasts about a month. Explanations for Earth, Moon, and Sun phenomena, such as phases of the Moon and eclipses, should be observational at this age, not explanatory.

Middle School Students

In the middle grades, students develop a more sophisticated understanding of the Earth, Moon, and Sun system, including explanations for the phases of the Moon and eclipses. At this grade level, they should be able to compare and contrast the difference between a lunar and solar eclipse as well as the difference between what causes a lunar eclipse versus the phases of the Moon. The use of models at this grade level is very important for developing an understanding of Earth-Moon-Sun phenomena.

High School Students

Students at the high school level expand their understanding of lunar eclipses to explain the different types of lunar eclipses, depending on where the Moon is in relation to the Earth's shadow. Their increased understanding of spatial geometry helps them visualize the conditions that must be in place in order for an eclipse to occur. However, many high school students are still confused by Earth-Moon-Sun phenomena like eclipses and phases of the Moon.

Administering the Probe

Ask students if they have ever experienced a lunar eclipse. Consider showing a photograph of a lunar eclipse from the internet or other source when using this probe, as long as it doesn't clue students as to the correct answer.

Related Ideas in *National Science Education Standards* (NRC 1996)

K–4 Changes in the Earth and Sky

* Objects in the sky have patterns of movement. The Moon moves across the sky on a daily basis much like the Sun. The observable shape of the Moon changes from day to day in a cycle that lasts about a month.

5–8 Earth in the Solar System

★ Most objects in the solar system are in regular and predictable motion. Those motions explain such phenomena as the phases of the Moon and eclipses.

Related Ideas in *Benchmarks for Science Literacy* (AAAS 1993 and 2008)

Note: Benchmarks revised in 2008 are indicated by (R). New benchmarks added in 2008 are indicated by (N).

★ Indicates a strong match between the ideas elicited by the probe and a national standard's learning goal.

K–2 The Universe

- The Moon looks a little different every day but looks the same again about every four weeks.

3–5 Motion

- Light travels and tends to maintain its direction of motion until it interacts with an object or material. Light can be absorbed, redirected, bounced back, or allowed to pass through. (N)

6–8 The Earth

- The Moon's orbit around the Earth once in about 28 days changes what part of the Moon is lighted by the Sun and how much of that part can be seen from the Earth. These changes constitute the phases of the Moon.

Related Research

- Phases of the Moon are often explained using the idea of a shadow of the Earth cast on the Moon, thus confusing eclipse phenomena with Moon phases (Driver et al. 1994).

- Astronomical phenomena are very challenging for students. To understand these phenomena, students must first understand how the Moon gets its light from the Sun as well as understand motion and position in the Earth-Moon-Sun system (AAAS 1993).

Suggestions for Instruction and Assessment

- A lunar eclipse is a phenomenon that is best explained and understood using physical models. However, merely observing a model demonstrated by the teacher is less effective in building understanding than having students manipulate their own models. It is important that students have opportunities to make and use their own models to understand eclipse phenomena.

- Have students use their models to compare and contrast a lunar eclipse with a new Moon. Comparing the two phenomena can be helpful in challenging their preconceptions about both events.

- Google NASA websites to find data on and activities related to lunar eclipses. There are many websites developed for education that target eclipse ideas.

- The PRISMS (Phenomena and Representations for Instruction of Science in Middle Schools) website, an NSDL (National Science Digital Library) resource, maintains a collection of reviewed lunar eclipse phenomena and representations teachers can use for instructional purposes. To access this collection, go to *http://prisms.mmsa.org*.

Related NSTA Science Store Publications, NSTA Journal Articles, NSTA SciGuides, NSTA SciPacks, and NSTA Science Objects

Association for the Advancement of Science (AAAS). 2001. *Atlas of science literacy.* Vol. 1. (See "The Solar System," pp. 44–45.) Washington, DC: AAAS.

Gilbert, S., and S. Ireton. 2003. *Understanding models in Earth and space science.* Arlington, VA: NSTA Press.

National Science Teachers Association (NSTA). 2008. Earth, Sun, and Moon. NSTA SciGuide. Online: *http://learningcenter.nsta.org/product_detail.aspx?id=10.2505/6/SCP-ESM.0.1*

Related Curriculum Topic Study Guide

(Keeley 2005)

"Earth, Moon, Sun System"

References

American Association for the Advancement of Science (AAAS). 1993. *Benchmarks for science literacy.* New York: Oxford University Press.

American Association for the Advancement of Science (AAAS). 2008. Benchmarks for science literacy online. *www.project2061.org/publications/bsl/online*

Driver, R., A. Squires, P. Rushworth, and V. Wood-Robinson. 1994. *Making sense of secondary science: Research into children's ideas.* London: RoutledgeFalmer.

Keeley, P. 2005. *Science curriculum topic study: Bridging the gap between standards and practice.* Thousand Oaks, CA: Corwin Press.

National Research Council (NRC). 1996. *National science education standards.* Washington, DC: National Academy Press.

Solar Eclipse

People have always been fascinated by solar eclipses. During a solar eclipse, parts of the Earth experience darkness for a brief time during the day. Throughout time, people have had different ideas about what happens during a solar eclipse. Here are some of their ideas:

A One of the nearby planets passes between the Sun and the Earth.

B The Sun passes between the Earth and the Moon.

C The Earth passes between the Sun and the Moon.

D The clouds block out the Sun.

E The Earth's shadow falls on the Sun.

F The Moon's shadow falls on the Earth.

G The Sun shuts off light for a few minutes.

H The Sun moves behind the Earth for a few minutes then comes back again.

Circle the letter of the idea that you think best explains what happens during a solar eclipse. Explain your thinking about solar eclipses.

_____

_____

_____

_____

_____

_____

_____

Solar Eclipse

Teacher Notes

Purpose

The purpose of this assessment probe is to elicit students' ideas about eclipses. The probe is designed to find out what students think happens during a solar eclipse.

Related Concepts

solar eclipse

Explanation

The best answer is F: The Moon's shadow falls on the Earth. A solar eclipse occurs during a new Moon when the Moon is directly aligned with the Earth and the Sun. A bright halo of the Sun's light (the solar corona) surrounding the disk of the Moon can be seen with the naked or unaided eye. Only during this brief time— when the Moon covers the Sun completely (for four to six minutes)—can the

Sun be viewed without special eye protection. Leading up to and after these brief moments of totality, special protective eclipse glasses must be worn or the Sun should be viewed through projection.

Because the Moon is blocking the light from the Sun, it causes a shadow to fall on the Earth, which results in the eerie change from daylight to darkness. To see a total eclipse, one must be in the path of totality. This path can be up to 200 miles wide and is the path that the Moon's shadow makes on the Earth during the eclipse. Observers within a certain distance away from totality can see a partial eclipse, in which the Moon does not completely obscure the Sun.

Curricular and Instructional Considerations

. . . .

Elementary Students

In the elementary grades, students learn basic ideas about the Moon and the Sun. Explanations for Earth, Moon, and Sun phenomena such as phases of the Moon and lunar and solar eclipses should be observational, not explanatory, at this age.

Middle School Students

In the middle grades, students develop a more sophisticated understanding of the Earth-Moon-Sun system, including explanations for the phases of the Moon and eclipses. At this grade level, they should be able to compare and contrast the difference between a lunar and solar eclipse and be able to state what causes a solar eclipse. The use of models at this grade level is very important for developing an understanding of Earth-Moon-Sun phenomena.

High School Students

Students at the high school level develop a more sophisticated understanding of eclipses by using their increased understanding of spatial geometry and understanding of light and shadows. Students can also use databases to look for patterns and to predict when and where an eclipse will occur. However, students at this level may still harbor commonly held misconceptions about eclipses.

Administering the Probe

Ask students if they or anyone they know has ever experienced a solar eclipse. Consider showing a photograph of a solar eclipse from the internet or other source when using this probe, being careful not to clue the students as to the cause of the eclipse.

Related Ideas in *National Science Education Standards* (NRC 1996)

. . . .

K–4 Changes in the Earth and Sky

- Objects in the sky have patterns of movement. The Moon moves across the sky on a daily basis much like the Sun.

5–8 Earth in the Solar System

- ★ Most objects in the solar system are in regular and predictable motion. Those motions explain phenomena such as the phases of the Moon and eclipses.

Related Ideas in *Benchmarks for Science Literacy* (AAAS 1993 and 2008)

. . . .

Note: Benchmarks revised in 2008 are indicated by (R). New benchmarks added in 2008 are indicated by (N).

K–2 The Universe

- The Sun can be seen only in the daytime, but the Moon can be seen sometimes at night and sometimes during the day. The Sun, Moon, and stars all appear to move slowly across the sky.

★ Indicates a strong match between the ideas elicited by the probe and a national standard's learning goal.

3–5 The Universe

• The Earth is one of several planets that orbit the Sun, and the Moon orbits around the Earth.

3–5 Motion

• Light travels and tends to maintain its direction of motion until it interacts with an object or material. Light can be absorbed, redirected, bounced back, or allowed to pass through. (N)

6–8 The Earth

• The Moon's orbit around the Earth once in about 28 days changes what part of the Moon is lighted by the Sun and how much of that part can be seen from the Earth— the phases of the Moon.

9–12 The Universe

• Mathematical models and computer simulations are used in studying evidence from many sources in order to form a scientific account of the universe.

Related Research

• Students' understanding of eclipse phenomena may be related to their lack of understanding of the relative sizes and distances apart of the Sun, Earth, and Moon. Many students draw these objects so they are the same size or between half and double each other's size. They also draw the Sun and Moon within one to four Earth diameters away from the Earth (Driver et al. 1994).

• Astronomical phenomena are very challenging for students. To understand these phenomena, students must first understand the motion and position in the Earth-Moon-Sun system (AAAS 1993).

Suggestions for Instruction and Assessment

• A solar eclipse is a phenomenon that is best explained and understood using physical models. However, having students merely observe a model demonstrated by the teacher is less effective in building understanding than having students manipulate their own models. It is important that students have opportunities to make and use their own models to understand eclipse phenomena.

• Have students use their models to compare and contrast a lunar and a solar eclipse.

• Google NASA websites to find data on and activities related to solar eclipses. There are many websites developed for education that target eclipse ideas.

• The PRISMS (Phenomena and Representations for Instruction of Science in Middle Schools) website, an NSDL (National Science Digital Library) resource, maintains a collection of reviewed lunar and solar eclipse phenomena and representations teachers can use for instructional purposes. To access this collection, go to *http://prisms.mmsa.org*.

Related NSTA Science Store Publications, NSTA Journal Articles, NSTA SciGuides, NSTA SciPacks, and NSTA Science Objects

Gilbert, S., and S. Ireton. 2003. *Understanding models in earth and space science.* Arlington, VA: NSTA Press.

National Science Teachers Association (NSTA). 2008. Earth, Sun, and Moon. NSTA SciPack. Online at *http://learningcenter.nsta.org/product_detail.aspx?id=10.2505/6/SCP-ESM.0.1*

Riddle, B. 2005. Science scope on the skies: You're blocking the light. *Science Scope* (Oct.): 70–72.

Riddle, B. 2007. Scope on the skies: Total lunar eclipse. *Science Scope* (Mar.): 76–78.

Related Curriculum Topic Study Guide

(Keeley 2005)

"Earth, Moon, Sun System"

References

American Association for the Advancement of Science (AAAS). 1993. *Benchmarks for science literacy.* New York: Oxford University Press.

American Association for the Advancement of Science (AAAS). 2008. Benchmarks for science literacy online. *www.project2061.org/publications/bsl/online*

Driver, R., A. Squires, P. Rushworth, and V. Wood-Robinson. 1994. *Making sense of secondary science: Research into children's ideas.* London: RoutledgeFalmer.

Keeley, P. 2005. *Science curriculum topic study: Bridging the gap between standards and practice.* Thousand Oaks, CA: Corwin Press.

National Research Council (NRC). 1996. *National science education standards.* Washington, DC: National Academy Press.

Index

Garrard, J., 129
gases, and "Burning Paper" probe, 24–25. *See also* greenhouse effect
genetics. *See* heredity; variation
geochemical cycles, 154
geography
    "Camping Trip" probe, 139
influences on students' thinking, xiii
"Where Would It Fall?" probe, 157–60
geology. *See* Earth science
germs and germ theory, 126, 127, 129
Gilbert, J., 58
"Global Warming" probe, 90, 143–49
grade levels, and use of probes, xi–xii. *See also* elementary students; high school students; middle school students
grading, of formative assessment, 5
gravity, 62, 70
greenhouse effect, 144, 147, 148

health
    "Catching a Cold" probe, 125–30
    "Is It Food?" probe, 94, 95
heat. *See also* heat energy; heat transfer; temperature
    "Magnets in Water" probe, 69
    "Moonlight" probe, 163
    use of term, 47
    "Warming Water" probe, 54
heat energy, 55, 56, 57
heat transfer, 138
heredity. *See also* evolution
    "Adaptation" probe, 116, 117
    "Is It Fitter?" probe, 122
high school students. *See* grade levels; *specific probes*
House (television program), 6
hydroxyl groups, 12

"Ice Cubes in a Bag" (Volume 1), 36
"Ice Water" probe, 10, 45–51
illustrations, as models, 74
immune system, 126
infectious disease, and "Catching a Cold" probe, 126, 128, 129
inquiry-based investigation
    "Burning Paper" probe, 28
    "Camping Trip" probe, 141
    "Standing on One Foot" probe, 64
    "Sugar Water" probe, 15
intent, and adaptation, 114
interdependence of life, 116
Intergovernmental Panel on Climate Change (IPCC), 148–49

ionic bond, 40
"Iron Bar" probe, 10, 17–22
"Is It Fitter?" probe, 90, 119–24
"Is It Food for Plants" (Volume 2), 96
"Is It Food?" probe, 90, 91–97
"Is It a Model?" probe, 10, 73–80
"Is It a System?" probe, 10, 81–87

journal articles, and NSTA publications, xiii–xiv

Keeley, Page, xiv
kinetic energy, 47, 54, 57
kinetic molecular theory, 18, 20

Lamarckian interpretations, 117
Lavoisier, Antoine, 28
Leader's Guide to Science Curriculum Topic Study: Designs, Tools, and Resources for Professional Learning, A (Mundry, Keeley, and Landel 2009), xiv
life. *See* origin of life
life cycles of organisms, and "Chicken Eggs" probe, 108
life science
    "Biological Evolution" probe, 99–104
    "Catching a Cold" probe, 125–30
    "Chicken Eggs" probe, 105–11
    concept matrix, 90
    "Digestive System" probe, 131–36
    "Is It Fitter?" probe, 119–24
    "Is It Food?" probe, 91–97
light
    "Magnets in Water" probe, 69
    "Moonlight" probe, 163
    reflection, 162
"Lunar Eclipse" probe, 90, 167–71

"Magnets in Water" probe, 10, 67–71
magnifiers, 42
magnetism, 68, 163
Maine Mathematics and Science Alliance, ix–x
Making Sense of Secondary Science: Research Into Children's Ideas (Driver et al. 1994), xii–xiii
mass, use of term, 26. *See also* conservation of mass
mathematical models, 74–75, 77, 78
matter. *See* conservation of matter; flow of matter and energy; phases of matter; properties of matter; structure of matter; transformation of matter
meteorites, 157
methane, 144
middle school students. *See* grade levels; PRISMS (Phenomena and Representations for Instruction of Science in Middle School); *specific probes*
minerals, 96